THE FORCES OF NATURE

THE FORCES OF NATURE

OUR QUEST TO CONQUER

THE PLANET

BARRY A. VANN

Prometheus Books

59 John Glenn Drive
Amherst, New York 14228–2119

Published 2012 by Prometheus Books

Cover image © Shutterstock.com/blinkblink
Cover design by Jacqueline Nasso Cooke

Inquiries should be addressed to
Prometheus Books
59 John Glenn Drive
Amherst, New York 14228–2119
VOICE: 716–691–0133
FAX: 716–691–0137
WWW.PROMETHEUSBOOKS.COM

16 15 14 13 12   5 4 3 2 1

Library of Congress Cataloging-in-Publication Data

Vann, Barry.
    The forces of nature : our quest to conquer the planet / by Barry A. Vann.
        p. cm.
    Includes bibliographical references and index.
    ISBN 978–1–61614–601–6 (cloth : alk. paper)
    ISBN 978–1–61614–602–3 (ebook)
    1. Human ecology. 2. Human geography. 3. Nature—Effect of human beings on.
4. Climatic changes. I. Title.

GF41.V6 2011
304.2—dc23

2011045847

Printed in the United States of America on acid-free paper

This book is dedicated to geographers Pete Charton, Harry Lane, Conrad T. Moore, and Chris Philo. My hope is that some of their fine legacy lives on in its pages.

CONTENTS

IMAGES

MAPS

TABLES

Chapter 1

THE POWER OF
ENVIRONMENTAL PERCEPTIONS

INTRODUCTION

Long before writers like ecologists Paul and Ann Ehrlich delved into
the causes and consequences of overpopulation, leaders in the Abrahamic religions taught that famine or food shortage was a sign of
God's wrath.[1] To think that people are just now starting to perceive the
world or its microspaces as overpopulated is not supported by human history. It is easy to imagine the fifty to sixty people that lived in a semipermanent Ice Age village debating among themselves about how to feed and
clothe their growing population. Being completely human, lasting disagreements sometimes occurred in those tight-knit communities. No doubt a
few disgruntled villagers would, from time to time, strap on their animal
hide shoes, pick up their flint spears, and leave the area in search of new
hunting and gathering grounds. Others who chose to stay behind benefited
from less competition for the wild edibles that kept them alive.

Even today, when the economy of an area becomes strapped, local
people must decide either to migrate or to stay and make changes in their
social relationships and culture. Overcrowded situations like these cause us
to depend on technology and complex social and economic systems to pack
ourselves into the places first settled by our ancestors ten to fifteen thousand years ago. Given that the earth is much bigger than the comparatively
small amount of land that serves as our living space or ecumene, it makes
sense to ask why we live on a relatively small and increasingly overcrowded
part of it.

Whether in ancient or modern times, people have always wanted to live in places situated on or near water bodies where life is best supported. To some observers, this need is deeply felt and affects us in ways that border on the metaphysical. From such diverse and thoughtful people as President John F. Kennedy (1917–1963) to the English professor and writer Norman Maclean (1902–1990), the mystical draw or pull that aqueous places have on the human heart have been poetically described. In his semi-autobiographical novella, Maclean wrote: "The river was cut by the world's great flood and runs over rocks from the basement of time. On some of those rocks are timeless raindrops. Under the rocks are the words, and some of the words are theirs. I am haunted by waters."[2] On the heels of the America's Cup races at Newport, Rhode Island, in September 1962, President Kennedy delivered a speech in which he claimed that "all of us have in our veins the exact same percentage of salt in our blood that exists in the ocean, and, therefore, we have salt in our blood, in our sweat, in our tears. We are tied to the ocean. And when we go back to the sea—whether it is to sail or to watch it—we are going back from whence we came."[3] Not withstanding these philosophical reflections on the deeper meaning of the human-water relationship, most authors who write about humanity's existence and quality of life think rationally in terms of food and protection from adverse environmental conditions. While no one can argue against the human drive to fill our bellies with life-giving grains and the flesh of foul, fish, and cattle, there is nonetheless a delicate dance between living near our source of fluids that carries oxygen and nutrients through our veins and our own demise. As *The Forces of Nature* shows, humans have historically sought home sites on or near water bodies in low-lying places, even if those places are situated in environmentally dangerous areas. The inappropriate use of technologies and the capricious nature of weather and climate can make life near seas, rivers, and lakes precarious, even deadly. Still, the immediate need to feed oneself and family is often perceived as the most pressing issue facing humanity's survival; so being situated near sources of flowing water helps insure that environmental conditions are ideal to support the growth of food items. Also, despite the changeable nature of weather near large water bodies, the actual range of annual temperatures is far greater in continental locations. With precious few uninhabited sites in more desirable midlatitude locations (between 30° north and 60° north) offering these advantages left uninhabited, we will increasingly need to adapt to less desirable and even inhospitable places. Out of necessity, we will depend more and more on technology to sustain our numbers.

Given the lack of desirable places left undeveloped and heavily settled, it is not surprising that developed countries in those temperate latitudes have adapted by reducing their growth rates; countries like Italy and Germany are actually declining. However, their relatively high standards of living attract immigrants from poorer regions with surplus populations. As long as this international "release valve" on population pressures exists, high rates of growth will not change in poorer countries. There will be no reason for them to adapt to limited resources, as they can simply relocate to places with greater carrying capacities.

A number of observers, including Paul and Ann Ehrlich, who contemplate overpopulation often begin with the insights of Rev. Thomas Malthus (1766–1834). His book from 1798 titled *An Essay on the Principle of Population* brought attention to the widening gap between food production capabilities and exponential population growth.[4] What is often missed from discussions about Malthus's work is the imagined world that existed in his mind. While most authors look at his notions of food production and population growth, which forms the famous J Curve, they often do not consider the fact that he was witnessing drastic social and cultural change in his immediate world, Great Britain and maybe western Europe, while outside of southern and East Asia, other parts of the world had much lower population densities. The urban population growth he witnessed was the result of the early stages of the Industrial Revolution that was born north of his home, in Scotland, during the preceding one hundred years.[5] One of the outcomes of this urbanization that Malthus did not foresee was the leveling off and even shrinkage of highly developed societies over time. Such is the case in countries like Japan, Germany, and Italy where the crude death rates now exceed the crude birth rates. Indeed, urban places have come to depend heavily on a continual supply of immigrants from highly fertile areas of the world.

Nonetheless, famine, disease, and warfare, in Malthus's way of thinking, were the expected outcomes of human life if moral restraint was not used in the area of human reproduction. An underlying reason for warfare, according to his ideas about overpopulation, is competition for resources. When people are hungry and there is little food to go around, they will, like most animals, resort to violence or conflict to stave off the pains of starving to death. A good number of people who have been confronted with this situation throughout history have chosen instead to migrate in search of sustenance and, hence, a more stable existence. When innovation, migration, and trade are not practical options for hungry people, food taboos are jettisoned, especially when starvation seems imminent. Look no further than to the Jamestown colony or the Donner party for examples of how humans will resort even to cannibalism to feed themselves and their families.

As mentioned previously, migration has been a principle means of coping with living in perceived overcrowded conditions. As a case in point, consider the deeper significance of what occurred at L'Anse aux Meadows in Newfoundland between 1002 and 1003. According to Greenland sagas and meager archeological evidence uncovered in that grassy place, there is little doubt that a family reunion occurred, although the participants probably did not see the meeting as a happy family gathering. The European migrants and the Natives they encountered were descendants of a man who had lived somewhere in central Asia only forty thousand years earlier.[6] That statement, however, may seem odd without a woman to serve as their ancestral mother. If one considers that a powerful man such as a clan or tribal chief often had more than one mate with lots of offspring, called the Genghis Khan (ca. 1162–1227) effect, it is not peculiar. Indeed, it has been estimated that there are sixteen million men alive today who carry Khan's Y chromosome.[7] Also, the unbroken line of maternal descent as observed in mitochondrial DNA does not lead to his time and place because tribal chiefs and other powerful men like Khan claimed so many women through whom they passed their respective Y chromosomes that geneticists have to go much farther back in time to find a common female ancestor. Nevertheless, the epic journey that the descendants of this man had completed covered thousands of miles, over treacherous and hostile lands and near-frozen seas. Their world was certainly not the idyllic Garden of Eden imagined by those who regard the pre–Industrial Age as such a place. On the other hand, their homeland was probably not a desolate wasteland either. This situation begs a fundamental question: Did the early ancestors of Asians, Pacific islanders, Native Americans, and Europeans move because of wanderlust or hunger? An even deeper question arises from a hunger-based scenario: Did these hunters and gatherers who populated parts of Europe, Asia, and the Americas move because of competition for food; or were they actually driven to migrate to find game and plants on which to feed because of imminent starvation? Given that they left others behind who continue to share their genetic sequences, it would seem that our ancestors regarded new horizons as opportunities to bag game and gather plants under conditions of less competition or simply to get away from others, including political leaders, with whom they disagreed on one or a host of issues. There is no way of knowing what prompted them to move. One thing is for certain, however: in a matter of a few thousand years, the descendants of the first modern people to leave northeast Africa populated the broad parameters of the human ecumene.

During three-fourths of the time span that elapsed between the family separation in central Asia and the meeting in Newfoundland, the world was beset by cold climate conditions. Sea levels were lower than they are today, and glaciers stood two miles above the earth's surface in places as far south as the modern American breadbasket in Kansas. As the climate warmed over the last ten to twelve thousand years, rising sea levels drowned well-worn trails across land bridges and flooded semipermanent, coastal settlements inhabited by hunters and gatherers. It was during a warming trend that ended before the fourteenth century that seafarers ventured out of Scandinavia and eventually across the North Atlantic to set up farms on the Shetlands and Orkneys, Hobrides, Ireland, Iceland, Greenland, and eventually Newfoundland at L'Anse aux Meadows. In much earlier times, perhaps at the end of the Pleistocene about twelve thousand years ago, hunters and gatherers traveling east used small boats to migrate along the shoreline of Asia to begin the human settlement of the Americas. Permanent contact between western migrants and their eastern counterparts did not happen until 1492 on the island of Hispaniola, when Christopher Columbus set foot on dry land.

It is interesting that Columbus perceived the seas and the world differently than most of his contemporaries. Although some fifteenth-century ship captains and traders who operated out of Atlantic and Mediterranean port cities believed that the world was round, they contended that it was too big to sail west to Asia. Columbus also reckoned that the world was round but that journeys to the spice markets of east Asia could be reached by setting sail westward. He had calculated the earth's circumference to be eighteen thousand miles, with Asia sitting only 3,500 miles east of the Canary Islands.[8] He was about seven thousand miles short in his estimate. When Columbus spotted land, he felt certain that his ships had arrived in India. The appearance of the people he met in the Caribbean certainly supported his conclusion, so he and others that followed him called the natives Indians. As geneticists Spencer Wells and Bryan Sykes have shown, Columbus was at least right about the genetic connections between the natives he encountered and their Asian cousins.[9] It is also interesting that Columbus had resurrected the geographic ideas of a long-dead Greek scholar named Eratosthenes, who combined the Greek words "geo" and "graphia" to form a new word and discipline he called geography. Despite living some sixteen hundred years later, Columbus's approximation of the earth's circumference was substantially shorter than that which Eratosthenes had estimated in the third century BCE. Amazingly, the calculation made by the ancient scholar was within 0.5 percent of the earth's actual circumference.[10]

Just like disagreeing Ice Age villagers and fifteenth-century ship captains, differing perceptions of environmental conditions have prompted discourse among geographers and even laypeople on the street. Upon closer examination, the argument that the environment sets insurmountable limits on us is no different than those same debates that have raged across time and space. It is easy to imagine that during the Ice Age, a group of, say, a dozen people stood on the shores of an icy sea and debated the merits of fashioning paddles to move along the coast in search of fish and seals. In a similar fashion, residents of New Orleans, for example, have disagreed over whether they should move to safer environs or if they should trust the technology that has allowed them to live in a place that is below sea level and virtually surrounded by water. Such has characterized life and discourse among humans since the dawn of our existence.

HISTORICAL PERSPECTIVES ON HUMAN-ENVIRONMENTAL INTERACTIONS

The study of past human-environmental interactions can be aided by combining paradigms that shape contemporary studies with the approaches employed by historical geographers. Using the three basic ways in which geographers examine historical places and events, as identified by geographer H. C. Darby, this study combines or overlays geographic paradigms with elements of Darby's "history behind geography," "past geographies," and, to a lesser extent, "geography behind history" approaches.[11] In the first case, Darby argued that historical geography can best be written by reconstructing past landscapes of human artifacts and activities that represent a temporal cross section with respect to a given spatial unit. His idea of history behind geography envisions tracing landscape impacts of long-term social and cultural developments of, say, ten or even hundreds of years. His third construct, which many historians and some geographers regard as "geographical history" is devoted to describing how past spatial features, including natural phenomena, have influenced historical events.[12] The line between what constitutes the work of historians and that of geographers is vague indeed.

Cambridge University professor Alan R. H. Baker discussed this disciplinary quandary in his *Geography and History: Bridging the Divide*.[13] Even Carl Sauer (1889–1975), who is often regarded as a historical geographer, described his own work as fitting into the corpus of a broader cultural history.[14] While many of the topics in this book are reconstructions of past geographies, some discussions involve events such as the hurricane that devastated Galveston, Texas, in September 1900. Clearly geography impacted history on that fateful day. Chapter 6 of this book, devoted to the Cumberland Gap area in the 1930s, clearly appeals to Darby's "history behind geography" approach. However, an approach to writing historical geography that seems to be lacking in the literature on these subjects is the superimposition of geographic paradigms onto Darby's approaches to the study of past places, events, and circumstances that clearly impacted landscapes that may or, indeed, may not continue to exist. The geosophy of this discipline is complicated, and its development is interesting and worth considering, for it reveals the inner workings of geographic thinking and, hence, environmental perceiving. As the overall progression of geographic thought reveals some of the paradigms that inform the case studies that follow, it is helpful to turn our attention to it.

For nearly two thousand years the work of writers such as Eratosthenes and Roman scholars Strabo and Ptolemy was primarily a tool for imperialistic regimes. The Greeks, Romans, and Ottoman empires were keen to know what lay on the other side of an unexplored mountain, sea, or desert. Geographers were employed as cartographers to draw maps and write exhaustive descriptions of people and lands. It was information compiled by geographers that told imperial policy makers where to draw boundaries or where to invade and conquer. Roman geographers, for instance, deemed Ireland too cold for the empire's purposes. In recognizing this environmental feature of Ireland, an unknown Roman bureaucrat named the island Hibernia, meaning "the land of winter." Roman contact with Ireland was henceforth limited to trade and the later evangelical work of fifth-century romanized monks, primarily Palladius in the south and Patrick in the north of the island. Nevertheless, geography was and continued to be an "encyclopedic discipline" until the mid-nineteenth century.

As products of the Enlightenment, European and American universities by the eighteenth and nineteenth centuries were teaching sciences along with traditional disciplines. The sciences offered greater insight into the problems facing civilization; more and more prestige was attached to those areas of study that promised answers to perplexing problems or a more efficient means to mine and fabricate natural resources into wealth. The German school of geography, led by Alexander von Humboldt (1769–1859) and Carl Ritter (1779–1859), sought to identify laws that handily explained life and culture. Since Eratosthenes sat down to describe the earth as he knew it, understanding and explaining human-environmental relationships has been a fundamental exercise of geographers. Unlike their predecessors, however, Humboldt and Ritter conceptualized geography as a science with nature determining all aspects of human biology and culture.

Members of the German school of geography were not the last contributors to the discipline's eclectic list of paradigms, yet their imprint is seen in contemporary debates over a host of topics including global warming. Nowadays, professional geographers and others interested in the products of this trade are witnessing an interesting intellectual era that nurtures two conflicting visions of the future and of discoveries yet to be made. On the one hand, there are those who see a bright future full of promise, which is made possible by innovative technologies. To them, humans possess an incredible capacity to overcome nature's limitations. They believe that through technological innovations virtually all things are possible. This human-empowering paradigm is handily known as *possiblism*. Its early proponent was Paul Vidal de La Blache (1845–1918), a French geographer. With such an optimistic name to inspire them, advocates of this paradigm believe that cities can be engineered better, swamps can be drained, fields can be irrigated, new treatments for dreaded diseases can be found, and humans will someday find a way to overcome even alien environments on distant planets. The wide variation of cultural characteristics found around the world in response to similar environmental pressures is a reflection of the human capacity for ingenuity to overcome natural barriers to survival. On the other hand, for many other people who share elements of the German school's vision of *environmental determinism*, images of the future are tarnished by a heightened concern for the consequences of ill-conceived human-environmental interactions that they argue result in increased global temperatures and other deleterious impacts on terrestrial and aquatic habitats. The latter vision seems to be gaining momentum in the media and political arenas.

In response to an international concern for the future of life as we know it, governments and international organizations, including some corporations, are adopting policies that promise to safeguard habitats, people, and infrastructure from hurricanes, tsunamis, acid rain, ground and surface water contamination, as well as threats from intergalactic objects such as meteorites and asteroids, for, as most primary school children know, nature killed the dinosaurs. That knowledge leads us logically to ask the questions: Can nature end our existence on spaceship Earth? In order for human civilization to survive, must we deliberately shape our regional cultures to cope with the forces of nature? According to environmental determinists, this process has naturally occurred throughout human existence. As we will see in subsequent chapters, this sentiment has resurfaced again in European and American societies, although, for reasons we will explore shortly, few geographers want to attach the environmental determinist label to the way they view human-environmental interactions, perhaps because it grants less power to people in shaping their futures.

There is a particularly intense and widespread fear of the consequences of global warming or climate change. In some of the proposed and enacted legislations, such as the Kyoto Protocol produced by the United Nations Framework Convention, emerging policies are designed to reign in various spill-over costs associated with fixing environmental problems resulting from unbridled and thus uncontrolled capitalism. Those who hold a more optimistic view of the future often see actions such as these as red meat for the media's hungry news hounds that are keen to sniff out sensational stories for the public's consumption. Nonetheless, the logic behind government actions and popular press coverage is thus based on the notion that nature is telling us to stop, drastically curtail, or modify the manner in which we exploit the earth and its resources. In short, we must seek out new ways to live in harmony with the forces of nature that, if we fail to respond effectively, will spell our doom. Sustainability, which is another way of expressing the need for living in a state of environmental harmony, is the key theme that guides some of the developed world's emerging economic and environmental policies. To the chagrin of many, however, the United States has been slow to respond to lobbyists' demands for action to cool a planet made red hot by the exploitive actions of humanity. We will return to perceptions of climatic conditions in a later chapter, but for now, let us return to describing geographic paradigms that frame environmental issues discussed in the pages ahead.

As I previously mentioned briefly, the notion that the environment influences and even determines the cultures of people, which would be the obvious outcome of new legislation designed to curtail environmental impacts, is an idea that many geographers cast aside decades ago. Among the turn-of-the-century proponents of environmental determinism to be cast aside in favor of the followers of possiblism were German scholars Humboldt, Ritter, and Friedrich Ratzel (1844–1904).[15] They, especially Ratzel, influenced American geographers Ellen Churchill Semple (1863–1932) and Ellsworth Huntington (1876–1947).[16] Their insights, too, were shunned by mid- and late-twentieth-century geographers.

While some of their more ambitious conclusions about human and environmental interactions have lost some of their currency in today's intellectual and politically correct marketplace, their descriptions of past environments ("past geographies") and their contribution to the development of the discipline are indispensable. Certainly, possiblism came along during a time in which there were rapid changes taking place in industry and on the farm. From home appliances to preserve and prepare food, to heavy machinery to drain wet places or irrigate dry lands, America's increasingly urban society had great faith in ingenuity. The natural world presented a mere challenge to a country that was annually adding millions of mouths to feed. However, the civil rights struggle, political assassinations, and Vietnam, along with other events in the 1950s and 1960s, ushered in a more realistic and perhaps even pessimistic outlook on American life. Simultaneously, an increased emphasis on science education that was prompted by America's participation in World War II and the subsequent "space race" against the Soviet Union increased society's understanding of science and its tangential application to environmental issues.

The stage was set for a so-called environmental reawakening. It was sparked in large part by the thought-provoking writings of American biologist Rachel Carson and environmentalist Aldo Leopold.[17] These popular writers were not alone in raising concerns about deteriorating environmental conditions. René Dubos's *Man Adapting* (1965), Donald Carr's *The Breath of Life* (1965), and William Wise's *Killer Smog* (1968), among others, stirred public concern for air, land, and water quality.[18] Many of the environmental laws passed in the 1970s as well as the creation of the Environmental Protection Agency (EPA) were products of these earlier movements and writings.[19] Clearly, Carson and Leopold, as well as those who have come after them, see limitations to the actions that humans may take with respect to interacting with the environment. More recently biologists Paul and Ann Ehrlich have carried the torch that was relit by Carson and Leopold's writings.[20] This reawakening, absent the broad conclusions offered by early environmental determinists, is nevertheless reminiscent of their underlying quest to understand how humans cope with the forces of nature.

Huntington, Semple, and their German predecessors were certainly products of their times. The prevailing views of most scholars who regarded themselves as social scientists were borrowed from their coworkers in the more respectable natural sciences. They reasoned that there must be laws that govern the social world, just as there are in the natural world. Perhaps no supposition appealed to these social scientists more than the theory of evolution advanced by Charles Darwin, especially his ideas on species' adaptations through natural selection and environmental conditions made famous in *On the Origin of Species* released in 1859. Darwin's ideas on competition within species and the corresponding notion of the survival of the fittest laid the groundwork for explaining global inequalities in wealth and culture. Sociologists, like geographers, borrowed the theory and explained macroeconomic patterns and other cultural variations with an evolutionary model handily called *social Darwinism*. In an age of striking global inequalities, some scholars and, unfortunately, political leaders latched on to the idea, as it seemed to explain Anglo-American and European achievements in medicine, transportation, military prowess, and institutionalized morality. The paradigm encouraged an already existing European ethnocentrism and seemed to grant license to political leaders to subjugate and rule over "lesser" peoples who were thought to be less evolved.

Geographers in the middle of the nineteenth century, unlike sociologists and historians, saw the environment as the primary reason for explaining cultural, including economic, patterns around the world. Ellsworth Huntington certainly helped to advance and ironically dismantle environmental determinism. Along the way, he and others like Paul Vidal de La Blache and Carl Sauer were instrumental in making geography apply scientific rigor, so the discipline would make conclusions based on systematic evaluations of evidence.[21] Of the early environmental determinists previously mentioned, Huntington is arguably the most important in this regard because he was instrumental in helping the discipline embrace scientific rigor and new perspectives. The geography he left behind was richer than the exploring and encyclopedic-like discipline he entered at the beginning of his career. No doubt the use of concepts and terms borrowed from biology, especially ecology, gave the discipline an academic respectability that added to its social prestige in academic halls, especially those located in major European and North American cities.

While Ellen Churchill Semple wrote about American issues, Huntington wrote mostly about the Mediterranean basin and Mesopotamia, so Americans are perhaps more familiar with Semple's work than that of Huntington.[22] Nonetheless, much is owed to Huntington, who labored as a professor at Euphrates College in Turkey from 1897 to 1901 and at Yale University from 1907 to 1917. His contribution to the field, however, must be framed in the historical context in which he lived and the disciplinary giants on whose shoulders he stood. If scholarly production is any indication of a person's love and dedication to a discipline, then Ellsworth Huntington had few peers. He and his American colleague Semple, like Alexander von Humboldt before them, wanted to make geography a respectable science. Huntington, at least, was keenly sensitive to the criticisms historians threw at his approach to geography. As his writings show, he responded to those critiques positively and used them to embrace more empirical ways on which to base cause and effect. His evolving awareness and its impact on his perceptions signaled a major departure from the encyclopedic conception of geography established in Greece during the third century BCE.

Academic geographers have not treated the legacy of environmental determinists or the work of Ellsworth Huntington well. This is perhaps due to the German school of geography's assumed association with early forms of the racial ideology that later developed in the minds of Nazi propagandists. Those ideologues used social Darwinism as the justification for speeding up the evolution of human society through the subjugation and genocide of so-called non-Aryans. That is an unfortunate and inappropriate condemnation of the entire work of environmental determinists. Still, the bias against their work remains a feature of the prevailing views of many geographers. Sadly, a review of recently published geography textbooks and specialized works on historical geography shows that information about Huntington has been nearly purged from their pages.[23] When he is mentioned, Huntington is given a secondary place among the likes of Semple, Ritter, Humboldt, and even Thomas Malthus. This is unfortunate on two fronts: First, his lengthy list of publications includes dozens of refereed articles and several books that, because they span the decades in which geography's favored paradigm changed from environmental determinism to possiblism and then into the cultural landscape approach made famous by a younger Carl Sauer, the corpus of his work provides us with a systematic description of the thought processes that went behind those disciplinary machinations. In short, they reveal much about the discipline's history and its development. Second, his descriptions of places located in the Mediterranean basin provide us with vivid depictions of past geographies, and that exercise, less some of his sweeping conclusions, continues to be a respectable activity for historical geographers.[24]

Ironically, some of the arguments made by environmental determinists like Huntington are reemerging in contemporary debates about global warming. Nevertheless, Huntington's thoughts, as reflected in his writings, reveal a man who wanted to use science to make the world a better place. He summed up his hopeful philosophy in the middle of his writing career:

> With the friendly help and criticism of our sister sciences we are advancing not only the time when geographical environment will be recognized as one of the great fundamental factors in the evolution of human character, but toward the time when we shall be able to show the processes by which it plays its part. When that is accomplished . . . knowledge of geographical laws will be of primary importance in enabling us to discover how certain evil traits of character may be eradicated and good ones fostered in their stead. Geography will be so interwoven with history that the two will be inseparable.[25]

It is anyone's guess what he meant by eradicating evil traits of character. Was the eradication to occur through an epiphany made possible through learning or through the tyrannical apparatus of government? Again, no one knows what Huntington saw as the vehicle for this kind of change, but there can be little doubt that a good many people perceive the loss of a benevolent, pristine world as justification for empowering government to rid society of the evil traits that make humans spoilers of the good earth. Although geography and history remain somewhat disconnected, his goals for social change through education, which were similar to the hopes and aspirations of Auguste Comte (1798–1857) who felt that all metaphysical faiths would give way to a social religion that he named sociology, are still with us.

Geographers have also considered another paradigm that Harlan Barrows (1877–1960) called human ecology.[26] In some respects, his school of thought is a compromise between environmental determinism and possiblism. Because of their argument that humans must adapt to environmental conditions, Paul and Ann Ehrlich's views fit neatly into this paradigm. However, their views are much closer to environmental determinism than they are to possiblism. Human ecologists who are philosophically close to possiblism argue that by expanding the earth's carrying capacity via economic structures and other modifications in material cultures, people will adapt to the limits imposed on them by nature.[27] However, human ecology has attracted criticisms from other geographers. According to Alan R. H. Baker, the renowned historical geographer from Cambridge University, Barrows's human ecology perspective in geography was highly criticized for its alleged naïve analogies, crude empiricism, and functionalist inductivism.[28]

Human ecology was a product of the progressive academic environment found at the University of Chicago in the 1920s and 1930s, and, in spite of Baker's limited appraisal of the paradigm, human ecology still garners respect from some biologists like the Ehrlichs and urban sociologists, including Herbert Gans, Walter Firey, and William Schwab. However, human ecology, as previously suggested with respect to its similarities to environmental determinism or possiblism, has mutated into two more subparadigms: the sociocultural approach and the neoorthodox approach.[29] While the sociocultural model emphasizes the role played by sentiments attached to family, political affiliation, and expressions of culture, ecological factors still influence people's choices in many public behaviors, including where they choose to live. For instance, a number of inner-city areas across the United States have retained high property values while in other urban areas, inner-city neighborhoods have deteriorated. The neoorthodox approach recognizes that problems plagued the paradigm's simplifications, especially those that argued that geographic communities were shaped by local environmental factors alone. Nonetheless, advocates of the neoorthodox approach still argue that technological innovations are nothing more than examples of ways in which humans adapt to various environments.

In summary and absent the role of divine forces, which is a perceptional topic discussed shortly, geographers see human-environmental relationships in one of three basic ways: the environment sets insurmountable limits on individuals and societies; humans can overcome environmental obstacles through innovative technologies; or humans adapt to environmental circumstances. As stated earlier, the third perspective is something of a compromise in that it recognizes that the environment does indeed present obstacles or limits, but that humans have historically found ways to live within nature's parameters. With human history as my guide, it is the contention of this author that societies naturally adapt to changing environmental conditions because the processes that inspire these offsetting perspectives have shaped all human societies, even those of our earliest ancestors and our nearest living cousins.

Even when humans make mistakes in trying to overcome an environmental obstacle, nature's resilience and ability to adapt to us helps to ensure the future of life. Consider the case of the kudzu, which is a fast-growing Japanese vine. As a nitrogen-fixing legume, the Soil Conservation Service (SCS) began a campaign in 1935 to plant the vine along American roadsides to control erosion. It flourished in the southeast, where weather patterns were ideal. As it turned out, erosion declined, but as an invader species, the kudzu vine has taken over spaces formerly occupied by native plants and was declared a pest weed by the United States Department of Agriculture in the early 1950s.[30] While not native to the southeast, and despite its classification as a weed, kudzu is now part of a vibrant regional ecosystem.

HUMAN VALUES AND
ENVIRONMENTAL PERCEPTION

Most societies have historically maintained religiously inspired images of lands, seas, nations, climate, and weather to help them cope with the forces of nature. Because of the infusion of divine imagery in shaping perspectives on spaces and the forces of nature that have accompanied human existence, it is necessary to include a perspective called *geotheology*. Inspiration and perspective on the early development of this important paradigm of human-environmental relationships can be drawn from E. G. Bowen, a Welsh historical geographer who held a strongly cultural, including religious, conception of spaces.[31] Celtic pagans and Christian religionists, he argued, have shaped Irish Sea coastal landscapes over the centuries. Other geographers are now recognizing religion's role in shaping spatial perspectives, but their approach falls short of truly entering the minds of those who have actually shaped the landscape, and along the way altered or created both secular and sacred spaces. In this area of inquiry, much has been accomplished by the likes of John K. Wright, Yi-Fu Tuan, Avihu Zakai, and, most recently, me—Barry A. Vann.[32] While delving into the imagined worlds of a few adherents to a religious system may suggest a comprehensive attitude toward nature, especially as they manifest themselves in regional and ethnic contexts, it is certainly not conclusive in the broader context of expansive, universalizing religions. There are some Christians, for example, who believe that the earth is available for exploitation and that people should be fruitful and multiply, while, on the other hand, other Christians see humanity as guardians of the planet and should therefore focus more on living in harmony with nature. It is interesting to note that these two attitudes produce behaviors that parallel those exhibited by proponents of environmental determinism and possiblism. It is also important to recognize that atheism does not necessarily mean that a person or society which holds that view is devoid of values and perspectives that fit nicely into environmental determinism, possiblism, cultural landscape, or human ecology paradigms. Indeed, nonreligious people can believe that nature rewards or punishes people for the quality of their environmental interactions.

The sacred texts of the three Abrahamic religions reveal environmental perceptions and imagery of climatic conditions in the Near East over a wide expanse of time. The Torah, Bible, and Qur'an each reveal images of spaces, including landscapes, seascapes, and weather events. Geographers like John K. Wright have developed a lexicon that can be used to place religious imagery into categories, which helps us understand how belief systems have influenced environmental perceptions and the ways in which people living in the cradle of civilization have coped with the forces of nature. These concepts would certainly be applicable to, for instance, Mesopotamian nature gods that predate monotheism.

As we will see in chapter 4, several geotheological terms are used that link together religious thought worlds with human-environmental relationships. Most of the terms and concepts employed in that chapter were coined by John K. Wright to identify the synthesis of religious and environmental geographies. The chapter's perspective on geotheology finds its roots in Wright's geosophy and R. W. Stump's identification of religious geography.[33] Specifically, terms such as geotheology (the general relationship between space and the worship of God); geopiety (religious, emotional attachment to space); geoteleology (the role of space in the unfolding of Providence); and geoeschatology (the role of space in the outcomes of Providence) are all significant in understanding religious imagery associated with spaces, including nations.[34] Wright's approach and his useful vocabulary call upon us to consider such interior imaginings in geographical research because of the role that imagings, feelings, and thoughts have in molding human-environmental perceptions and behaviors which often impact the landscape with visible patterns and structures.[35] As is shown in chapter 4, geotheology, as conceived by Wright, fails to capture the full range of the ways in which people associate the forces of nature with the divine; so new terminology is introduced that captures the belief that the forces of nature are used by deities to punish or reward individuals, towns, and nations. These new terms are further modified to identify ways in which people who exclude God from human-environmental relationships maintain that there are nonetheless negative consequences for living environmentally exploitative lifestyles and positive gains for living in harmony with nature.

THE IMPORTANCE OF
ENVIRONMENTAL PERCEPTION

The purpose of *The Forces of Nature* is to shed light on the ways in which people, especially in the United States, have interacted with the environment in past times, particularly those that involved mass population flows or deaths with implications for the future. One could arguably include hundreds of issues in a book of this kind, but it is limited to forces that have shaped the spatial patterns of contemporary human societies. While early in the book, attention is given to thoughts on how our ancestors shaped our ecumene, which illustrate human life in closest harmony with nature, the later chapters focus on the United States and the interplay between history, geography, and the consequences of those human-environmental interactions. It is my hope that the topics highlighted in *The Forces of Nature* will help illuminate past human-environmental interactions that have captured and perhaps even distorted the popular imagination. By appreciating past environmental events in the context of future population change, it is possible to see where real dangers await us in the future.

The goal of this book is to show how geographic paradigms, especially human ecology set in past times, can help frame contemporary environmental issues in a continuous temporal context. In this book, human ecology is conceptualized as something of a compromise between the human empowering paradigm known as possiblism and the opposing view of environmental determinism. Specific attention is given to these perspectives in the context of the Pattisonian man-land or human-environmental tradition.[36] In more practical terms, this suggests that while one geographer could approach a study on human-environmental interactions in the New Orleans basin in the late summer of 2005 from the perspective of a possiblist, another earnest researcher may be driven by an environmental determinist paradigm. Clearly the implications of their studies will be quite different. As a compromise, human ecology seeks to appreciate the ways in which microsocieties and macrosocieties adapt to environmental circumstances. It is the contention of this author that the human ecology paradigm offers the most objective perspective to examine precarious and life-threatening environmental situations.

SUMMARY

In the spirit of my older colleagues, the early sections of the book focus on the creation of the human ecumene and the establishment of ancient populations around the world. By doing this, we are better able to appreciate the environmental conditions that best support human life and sustainable communities. This discussion shows that, despite centuries of technological innovations to cope with the forces of nature, our species has strayed little from the migratory paths followed by our ancestors, the earliest *Homo sapiens*. Indeed as subsequent chapters show, the broad parameters of the ecumene that were marked by the earliest migrants from the Near East are now filling up and are influencing the ways in which people are adopting technology-dependent, urban lifestyles.

Before moving into the later chapters, some attention is given to geotheology, because humans in various places have attached spiritual meanings to spaces and the forces of nature. Geotheological imagings have certainly provided diverse peoples with a psychological framework for appreciating favorable environmental conditions as well as a way to connect human behaviors to adverse weather and tectonic events, which have been seen as dispensed by the hand of God. The three holy texts of the Abrahamic religions also provide insights into ancient and early medieval climatic conditions in the Near East. Imagery in those texts suggests that significant climate change has occurred in the Near East over the past 3,500 years. On a related topic is the notion of human-causation of negative environmental conditions such as climate change. To frame that evolving social reality, the concepts of geokolasis and geomisthosis are introduced, for even secular thinkers still connect human conduct with larger punishing forces involving nature. Indeed, the three Abrahamic texts reveal a deep-seated belief in geotheokolasis and geotheomisthosis, which mean literally "earth, god, punishment" and "earth, god, reward or payment," respectively. With or without God, it seems that we are a species intent on injecting ourselves into environmental matters.

Nevertheless, population pressure is increasingly forcing us to settle in less desirable and inhospitable places. Many of these seemingly benign homesteads are environmentally dangerous and invite human-environmental interactions that elicit geotheokolasis or geokolasis sentiments. One of the ways that people cope with the awareness of their precarious situation is faith in God or technology or a combination of both. Regardless of whether one believes that God is the author of the laws of nature, environmental forces are nevertheless directed by noncompromising decrees. The use of technology cannot overcome some of nature's forces.

The challenges we face in regard to using technology to cope with the forces of nature are presented in later chapters. Topics focus on the United States because, since the Industrial Revolution, well-meaning innovations designed to irrigate dry lands or drain wet areas have lured a significant number of people to settle in environmentally fragile and unfamiliar places. While most of these areas seem inviting, they are nonetheless situated in the crossroads of extremely powerful natural forces. When those forces collide, death and destruction are often the results. Recently established settlements with increasingly high population densities will, at some unknown time in the future, be hit by devastating droughts, earthquakes, floods, hurricanes, and/or tornadoes. Clearly there is a perceptual disconnect between environmental reality and decisions that have shaped Americans' settlement choices.

In its entirety, *The Forces of Nature* uses case studies of past human-environmental interactions as a means to look deeper into how humans have perceived and coped with natural forces. Armed with knowledge about these deadly events, we can cast an eye toward the future; it is imperative that we make good decisions about where we live and the dependability of available technology to protect us from the forces of nature.

Chapter 2

LIVING ON THE EDGE
OF THE WORLD

INTRODUCTION

As new developments are made in the fields of archeology, historical geography, and genetics, the mysteries of human prehistory are being revealed in amazing ways. What is striking is that the evidence suggests that our ancestors were gifted with astute geographic intuition. Their perceptions and accumulating knowledge of human-environmental interactions helped them to cross dangerous, unknown lands and even violent ocean swells during the Mesolithic and Neolithic ages. Driven by the hope of a better life made possible through less competition with others over sometimes meager food supplies, our ancestors were forced to move away from family and loved ones. Because geographic knowledge and perspective were matters of life and death, they had little choice but to carve out an ecumene (inhabited space) that continues to be our own as their descendents, for we still walk in their footsteps in many and often forgotten ways.

FORMING THE HUMAN ECUMENE

In contrast to the multinuclei theory of human origins, archeological evidence points to east-central Africa as the hearth or original homeland of modern humans. Despite some minor disagreements among geneticists, their work is lending support to the Out-of-Africa theory. Spencer Wells's research in genetics, for example, suggests that every living person on the planet is descended from a man who lived in an eastern African valley some sixty thousand years ago.[1] However, as Oxford University geneticist Bryan Sykes, historian Helen A. Gaudette, and paleoanthropologist G. A. Clark point out, modern people had already left Africa and were in Israel some forty thousand years earlier.[2] Mitochondrial DNA, which is passed from generation to generation along the female line, builds a case for a small band of emigrating Africans that included one woman, or her recent descendants, who was the maternal ancestor of all non-African peoples.[3] Nevertheless, and despite debates among scientists about the precise date for *Homo sapiens'* first migrations out of Africa, the way in which these earliest humans used geographic knowledge to cope with the forces of nature deserves appreciation.

Long before Eratosthenes, the ancient Greek scholar who coined the name geography to identify his work, our ancestors used their practical understanding of human-environmental interactions for their survival. For reasons that no one has determined, *Homo sapiens*, upon leaving their African abodes, lingered in what is today Israel and the Near East for fifty thousand years. The Pleistocene epoch (the last great ice age) and its continental glaciers were certainly intervening obstacles for northward migrations, as the southern boundaries of the glaciers prevented northern migration past the eastern Mediterranean. It is also likely that the area of Israel was a temperate land with lush meadows and green forests. Plentiful game made life easy for hunters. Given that the Out-of-Africa movement was isolated to the descendents of one woman during a time in which ice blanketed much of the upper-middle and high latitudes, the descendants of that one woman, whom geneticist Bryan Sykes named "Lara," begat the population of people who most likely used their time in that portion of the Near East to develop the technology to make living in the colder reaches of Europe and Asia possible.[4]

Interestingly enough, and almost simultaneously (in geological time), it took another thirty-eight thousand years for modern people to settle the farthest reaches of western Europe and the Americas.[5] The migration routes of those earliest explorers and pioneering migrants is well marked by archeological evidence deposited during the late Paleolithic age (up to 15,000 BP, or before the present) and Mesolithic era (15,000 to 5,000 BP).

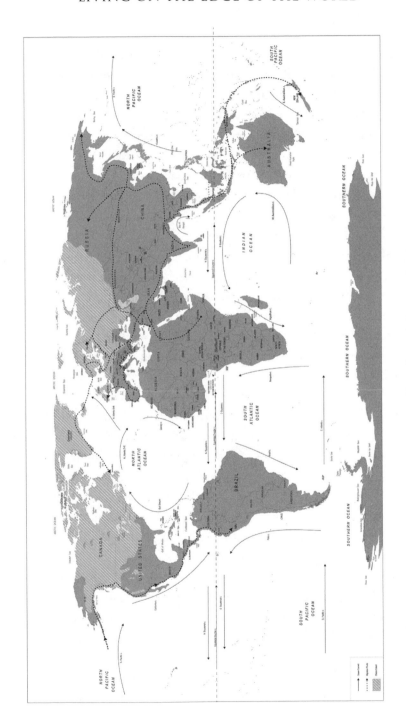

Map 2.1. Ice Age and Early Human Migration. *Image courtesy of Geraldine Allen and Clayton Andrew Long.*

Genetic material and artifacts left behind in caves and buried under peat and layers of other sediments tell us much about how they perceived, inter-acted, and responded to the environment. Their environmental perceptions and interactions are important for us to consider because the decisions they made in regard to where to live and what materials to use outlined, with minor revisions made possible through technological innovations over the last two hundred years, the boundaries of our ecumene.

We know for certain that through the Mesolithic Era, at least, most people were hunters and gatherers. As such, their survival depended heavily, if not entirely, on finding and following food supplies. Traveling in small bands of people, who were perhaps sometimes spurred on by curiosity about what lay over the next hill or mountain, they formed migration routes through valleys, plains, and mountain passes. Migration routes extended away from the Near East until land gave way to sea. Then, through ingenuity, some people constructed small boats and moved on, following coastlines where kelp beds rich with life served them well. As hunters and gatherers, the pursuit of game and plants that gave them their sustenance led them across new and unfamiliar lands. These Paleolithic and Mesolithic peoples left their stone tools and genetic signatures behind them as they established set-tlements on six of the earth's seven continents. In a geological time frame, humans have only claimed dominion over those six continents for a minis-cule amount of time, about twelve thousand years. While genes and artifacts make it possible to re-create their routes and the reasons for them, compar-isons to them with our present-day population clusters reveal the depth of their geosophy (geographic knowledge) and the sustainability of the ecumene they chose for themselves and their descendants.[6]

Since the 1980s, the work of geneticists has corroborated much of what archeologists, linguists, and historical geographers have suspected for decades. By using quite complex methods to analyze DNA sequences and mutations in them, it is possible to estimate the times that two groups or even individuals had been separated. This information, along with arche-ology as well as sea and land features, has allowed us to determine that there were two migration routes from Israel. One went north and then split into three directions that eventually stretched across Europe and Siberia. At least one of those northbound routes went through the Caucasus Mountains and then split again. Another of these paths headed west across northern Europe, and the other headed east across Asia and through the Americas.

The eastern main route out of the Near East followed the coast along the Persian Gulf and then split into two branches. The southeastern route followed the coast around the Indian subcontinent, into Southeast Asia, and made its way to Australia, New Zealand, and Polynesia. A number of people following this route made good use of boats and lower sea levels that exposed land areas. They even populated the Hawaiian Islands. Because the interglacial period (Holocene epoch) brought warmer weather, alpine and continental glaciers melted, and the runoff caused sea levels to rise. Many of the islands that those seafaring folk had followed to Hawaii are now submerged under warm, tropical waters. Genetic evidence shows that it was during this period of climatic warming and coastal flooding that the south Asian route shifted northeastward along the coast and the Aleutians. This route did not end at the Bering Strait. By 11,000 BCE, descendants of Asian hunters and gatherers had made their way to the coast of northern South America.[7]

The European flows followed either a northwestern trajectory across the North European Plain toward Scandinavia or a coastal route around Iberia and eventually into Ireland.[8] By the late ninth and tenth centuries, northern Europeans from Scandinavia had settled the North Atlantic's Faroe Islands, Iceland, and then Greenland. As was shown earlier, by the eleventh century, Leif Ericson and his followers were in Newfoundland. It is anyone's guess as to what happened to many of these westward wanderers. Were they assimilated into Asian populations who had already made it to North America? While runic stones suggesting an expansive Viking presence in North America have been proven to be hoaxes, written accounts of the Lewis and Clark Expedition that explored the upper Midwest and the Northern Rockies between 1803 and 1805 suggest that perhaps some of their genetic material could have been detected among the Mandan Indians. According to the explorers' accounts, these folk had light-colored hair and skin.[9] They did not, however, have immunity to European diseases, and many Mandans eventually succumbed to them.

Native Americans, including Eastern Woodlands Indians, were still migrating when Europeans arrived en masse in the sixteenth and seventeenth centuries. The Shawnee and Cherokee were each evolving cultural groups that split off from larger Algonquin and Iroquois peoples that populated southeastern Canada and what is today upstate New York. Shawnee arrived in southern Ohio in the mid-seventeenth century and Cherokee in southern Appalachia somewhere between 1000 CE and the mid-fifteenth century.[10] These two groups built semipermanent villages along waterways where they farmed, fished, and hunted to supplement their diets with protein.

By the end of the eighteenth century, environmental perceptions of nonprofessional geographers had led people to settle in places that today still mark the basic outlines of the human ecumene. It is an astonishing feat that our ancestors completed. In forty to sixty thousand years, they used the products and forces of nature to shape our habitat. Still, the forces of nature were formidable obstacles to overcome. Indeed, many people chose not to leave the best known areas in search of new places that others hoped were similarly endowed with life-sustaining features.

THE EDGE OF THE WORLD

On a rocky cliff overlooking the cold yet fertile waters of the North Atlantic, early Neolithic settlers, fresh off a short yet treacherous voyage across the choppy waters of the Pentland Firth, gazed out over the vast expanse of icy water before them and decided they had reached the edge of the world. Although there is no record of this event, it is assumed to have taken place because in 1850, a tempest struck the Orkneys and blew away sand dunes and their scant grassy cover, thus exposing a tiny hamlet that archaeologists named Skara Brae. It was a permanent settlement of perhaps fifty to sixty people living in sandstone houses with walls that protected residents' private sleeping and living quarters from neighbors and frequently hostile and deadly elements. Evidence found there reveals that these early Orcadians, as they have come to be known, enjoyed a well-balanced diet of corkfin wrasse, mussels, oysters, and red bream from the shallows. They cultivated wheat and barley and raised cattle for milk and meat.[11] This settlement may be crude by today's standards, but Skara Brae was established in Orkney long before the pyramids were built in Egypt and even two thousand years before later Neolithic people dragged Welsh rocks onto the Wiltshire plains to erect Stonehenge. At five thousand years old, Skara Brae is not, however, the oldest settlement found in Britain and Ireland; but like the oldest settlement, it is located close to the sea.

The oldest known settlement in Britain and Ireland is located at Mount Sandel in Northern Ireland, near the banks and mouth of the River Bann. It is a Mesolithic settlement that dates back to 7000 BCE, some four thousand years before Skara Brae was built.[12] A brief look at the Irish landscape suggests that the founders of Mount Sandel were quite astute. This part of Ireland is in the northeast, away from the gales that blow in from the Atlantic. The Irish Sea side of the island is, in an Irish sense, a gentler landscape. The ample supply of limestone rock neutralizes soil acidity and contributes to the lush green vegetation that gives Ireland its colloquial name: the Emerald Isle. It is little wonder that, supported by this fertile landscape, there are other Mesolithic sites in Ireland, including one at Sutton, but none as old as Mount Sandel.

Each of these places show that Mesolithic people built residential structures, but they also illustrate that these early humans had a concern for properly doing away with waste materials in landfills called *middens*. Archeologists have found therein bones discarded after hearty feasts on red deer, wild pig, salmon, shells from crustaceans, and, at a site on the Dingle Peninsula in Ireland, cattle bones. These waste dumps accumulated into mounds as much as five meters in height and one hundred meters in diameter. They are undeniably treasure troves from which archeologists can attempt to learn more about these ancient civilizations.

As evidenced by the remnants of feasts found in the middens, Mesolithic peoples understood that life on or near estuaries provided ample supplies of food. Imagine living in a place where one could regularly dine on fresh venison, pork, aurochs (ancestors of domesticated cattle), hazelnuts, wild pears, crab apples, and a variety of shellfish and other sea life. Because of decomposition, it is difficult to determine what other kinds of plant life supplemented their diets. Nevertheless, humans' preference for living by large water bodies is an ancient and common trait among all of our ancestors, no matter where they ventured. This brings to mind an observation made by the Welsh historical geographer E. G. Bowen who noted that at least since the Mesolithic culture period, coastal settlements functioned "both as places of refuge and as stepping-stones of coastal diffusion."[13] One might add that these locations were the most coveted spaces and may well have served as permanent living settlements, even before the Agricultural Revolution introduced permanent settlements in inland locations. This is seen in the types of materials found in middens. As Bryan Sykes points out, these sites, especially the one at Mount Sandel, were permanent enough for residents to use logs to construct houses as opposed to temporary lean-to structures. This is apparent from the location of postholes that would have served as insertion points for corner posts.[14] Among the seeds and bone debris found in the midden at Mount Sandel were animal and plant remains that suggest season-specific availability and consumption: "There are hundreds of salmon bones, which show that the site was occupied in the summer when salmon, fresh from the sea, pushed upstream to their spawning grounds. Huge numbers of hazelnuts and the seeds of water lilies, wild pear and crab apple show that the site was used during the autumn harvest of wild forest food. The remains of young pigs, which were born in late autumn, are the sure sign of winter occupation."[15]

Archeologists have also learned that the ancient residents used ringbarking to clear forests, which would allow for certain nut-bearing species like the common hazel (*Corylus avellana*) and perhaps other edible plants to grow. Although ringbarking is an effective way to bring down even a behemoth tree, it is a slow death for the plant. Nevertheless, it shows how Mesolithic people saw how different elements of nature could work together for their benefit. This suggests a people who were not mindlessly following herds of animals for their survival. Indeed, their use of ringbarking tells us that they understood tree physiology, although they would not have used words like phloem, cambium, and xylem cells to describe the intricacies of their work. By using stone-cutting tools, the ancients simply carved a ring around the tree, which effectively cut off the supply of food through the phloem or inner bark and then into the water-supplying pipeline formed by xylem cells in the outer wood.[16] Within a year or so, the tree would die, wither, and succumb to the ravages of a gale. The felling of trees allowed sunlight to penetrate newly opened spaces in the forest canopy, providing opportunities for new growth and browsing material for red and roe deer as well as new growth of hazel shrubs, the source of the resident's favored nuts. By promoting new plant and animal life in this way, early humans could supply their settlements with sustainable resources.

Fabricating felled trees into useful timber for building and cutting larger logs into smaller pieces for fuel to heat homes and for cooking in large pits required a reliable source of material suitable for cutting woody tissue as well as for butchering flesh and scrapping animal hides. Flint found in the River Bann area only added to the local environment's cornucopia of life-sustaining resources. Indeed, the river watershed provided local Mesolithic people with ample supplies of flint, buried in chalk layers. Archeologists have found flint from the Mount Sandel area in a number of other places. Their discovery certainly suggests that cultural diffusion occurred from Mount Sandel, but it may also show the extent of the hunting grounds of the semipermanent people who enjoyed life centered at Mount Sandel. Flint materials from this part of Ireland have been discovered down the east coast of Ireland, including in Dublin Bay and farther east, midway across the Irish Sea on *Ellan Vannin* (the Gaelic name for the Isle of Man). Flint from the River Bann area has also been discovered by archeologists along rivers and lakes headed southwestward into Connacht.[17] The evidence certainly hints that some Mesolithic peoples, at least at the edge of the European world, were living in a virtual Garden of Eden long before Neolithic farmers arrived from the Near East.

Ample evidence from coastal areas in east Asia show that early people made similar uses of such places. When Mount Sandel was flourishing, a similar village named Hemudu was built on an estuary in what is today eastern China. Archeological evidence found at Hemudu is dated to about 5000 BCE, about the same time that Skara Brae was established.[18] Located near the modern village of Hangzhou, the marshy coastland presented the ancient builders with some settlement challenges. No doubt the area was flooded from time to time, yet periodic flooding gave the land nutrients that fed native stands of wild rice, thus providing those who dared to inhabit the area with sustenance off of which they could thrive.

Hemudu and Mount Sandel are located at opposite ends of the earth's largest landmass (Eurasia), but, in the larger time period of human existence, the similarity in their ages is remarkable. These villages were not products of an agrarian society. They were built by people who clearly understood a great deal about their environments and chose sites that offered similar resources. Given that genetic research shows that Europeans and Asians share common ancestors who lived in central Asia about forty thousand years ago, it is almost as if their offspring had raced away from each other in opposite directions. If building a village at the edge of the world was their goal, it would be difficult to declare a winner—their ages are that similar to each other.

What is unfortunate for modern scholars interested in studying coastal settlements that likely existed earlier in the Paleolithic era and the early Mesolithic age is that rising sea levels over the last twelve to fifteen thousand years have hidden their treasures. If those places were still imprints on dry land, the spatial extent of the modern ecumene would need to be redrawn to include them. There can be little doubt that Paleolithic and early Mesolithic peoples chose their homes or campsites well. Nevertheless, and despite the lack of archeological evidence from those earlier times, humans determined long before Eratosthenes lived and founded the discipline of geography that coastal areas, especially estuaries, were optimum places to live, especially if those places were on the leeward side of land masses. As the next section of this chapter shows, modern humans have certainly maintained a similar perception of land near seas, lakes, and rivers.

RURAL TO RURAL MIGRATION

Scholars have long held that the Agricultural Revolution and the Neolithic age were born in Mesopotamia (land between the Tigris and Euphrates Rivers) around 8000 BCE. Planting predictable crops and raising cattle, sheep, goats, and horses gave humans a reason to establish permanent settlements. Shortly thereafter, agricultural practices were carried by migrants to other lands. It is also possible that trade and modest communication lines stretched along migration routes helped diffuse agriculture. However, as shown in the previous section, sedentism predates the practice of intensive agriculture. This was certainly the case in Mesolithic villages like Mount Sandel and Hemudu. Semipermanent settlements established during the Mesolithic age continued to attract others until their carrying capacity was reached. Early on, villages brimming with people typically sent out groups to colonize similar places, but when that option was no longer the most efficient means of coping with having too many mouths to feed, villages developed ways to domesticate local plants and animals. The accumulation of knowledge about the animals and plants on which they depended, as well as ideas picked up by trading with other villages, helped usher in agriculture.

In Europe, genetics and archeology shows that later newcomers from the Near East brought farming knowledge with them. The Neolithic age replaced the older way of life. Now, farming enabled people to move farther away from the copious shallows of coastal settlements and placid lakeshores, but rivers and streams, as sources of fresh water and fish, were always attractive places to build a village. Archeological evidence shows that the older Mesolithic peoples had few conflicts with their new neighbors. According to geneticist Sykes, it is more likely that Mesolithic and Neolithic peoples traded goods with each other and, as shown by genetics, certainly interbred.[19] Eventually, the distance between inland farms and coastal settlements diminished as each population grew toward each other. Over time, powerful invaders in Britain and Ireland used these coastal villages as ports and residences, forming cities like Dublin, Waterford, Wexford, and London. While the Vikings built the former, the Romans established London on the Thames as a major port of entry for their ships, armies, and merchants.

These Iron Age conquerors rather quickly dominated the inhabitants of southern Britain, but in other parts of the world, including sub-Saharan Africa and the Americas, open spaces allowed hunting and gathering to continue as a way of life as populations continued to move across uninhabited spaces. As nature provided our hunting and gathering ancestors with abundant food supplies and comparatively warmer weather, there was little need for them to change. Indeed, a similar process of cultural conservation is detectable among Irish and Scottish immigrants arriving in America's Colonial backcountry during the eighteenth century. These immigrants created a rural to rural migration flow across the Upland South; their agrarian lifestyle changed little until migration ceased to offer opportunities to continue with their way of life. This fact partly explains the existence of comparatively poor economic conditions found in more remote areas of southern Appalachia and the Ozarks.[20] Through westward expansion, the environment continued to supply them with ample food and water. Like the Native Americans before them, there was little reason for them to change their culture.

This and other situations mitigate perspectives argued by later geographers including Ernest George (E. G.) Ravenstein,[21] who, beginning in the late 1880s, have maintained that people migrating in large numbers follow a rural to urban pattern, but the evidence suggests that this pattern is found in history only after the ecumene, or parts of it, was established and agriculture as well as urban life became relatively appealing. More recently, Wilbur Zelinsky and Huw Jones developed mobility transition models arguing that overpopulation causes migration to occur in large numbers.[22] This theory of mobility transition and similar migration assumptions, like the gravity model, are relevant in the context of modern societies. How can the gravity model, which declares that larger settlements attract people from greater distances than small villages, apply to people who lived in an age in which large settlements did not exist? It is the contention of this author that natural resources, including plants, animals, climate, and rock material, attracted Mesolithic and early Neolithic peoples—not the diversity of urban life. Those modern theories consequently offer little to our understanding of how the ecumene was formed during the Mesolithic and early Neolithic eras. Fortunately, sociologists and anthropologists with interests in ancient cities and settlements offer insights into how that process likely unfolded.

Interestingly, some modern social theorists such as urban sociologist Bill Schwab developed the view that overpopulation during the Mesolithic era encouraged the rise of agriculture and the birth of cities, clearly requiring a sedentary population. As Schwab points out, the work of archeologists and anthropologists suggests that agriculture is an undesirable alternative to hunting and gathering, which, as a way of life, requires less labor than farming and caring for cattle, camels, horses, and sheep. Studies conducted among the few remaining hunting and gathering societies demonstrate that leisure activities abound; there are surpluses of time and personal energy because the people do not have to work very hard.[23] This interpretation for the rise of clustered settlements is a mirror reflection of the author's notion that many preindustrial migrations, which followed a rural to rural pattern, were done for survival reasons. People simply moved where there was less competition for the flora and fauna that sustained them.

BUILDING ON THE OLD

It may seem odd that people who have become geographically isolated from others whom their ancestors knew and were perhaps related to by blood can forget them and yet retain immaterial and material aspects of their shared culture. Those kernels of culture can sometimes be found across wide expanses of space. Accumulated knowledge of flora and inanimate environmental conditions most likely led to the Agricultural Revolution. For people on the move, retaining culture and ideas can help them build a new life on their ancestors' old ways. An international teaching and research experience early in my career made this part of rural to rural migration clear.

I was fortunate to hold an honorary teaching fellowship at the University of Dundee in Scotland. While conducting my first lecture in a course called Scotland and the Americas, a student asked if white Americans came from Europe. Upon hearing the question, I surveyed the room looking at the expressions of the fifty plus faces. I fully expected to see smiles of incredulity everywhere, but instead, I saw looks of curiosity as the students sat motionless, waiting for my answer. In thinking of an answer to the student's question, I remembered a situation that occurred in Oklahoma a few years earlier, when a waitress in a café asked the folks sitting at my table a question. She asked, "Are yunz ready to order?" I recalled that after hearing her query, I blurted out, "You said 'yunz,'" which was a word commonly used in my part of Southern Appalachia, to which she replied: "Yes, that's an Okie word." Taking that event in Oklahoma as an opportunity to use in a teachable moment, I returned to the Scottish student who had asked the question. "Well, yes," I said. "White people in America have roots in Europe." I then told the class about my experience in Oklahoma, and the lady's use of the word "yunz." This had relevance to them because "yunz" and similar pronunciations like "yinz" and "hankering" were used in Scotland during the time of Scottish migrations to Ireland. Later migration flows over the Atlantic Ocean carried the word across Pennsylvania and the American South and into rural Oklahoma where my waitress defended the word as quintessentially Oklahoman in origin. Both the students and the waitress shared at least one cultural trait, but they had no idea others living across the Atlantic shared it with them. Clearly the waitress had no idea that the word was commonly used in eastern Tennessee and eastern Kentucky. As this situation shows, people taking part in rural to rural relocations, as was certainly the case among hunters and gatherers and even subsistence farmers, do not often have strong communication ties with the old country. Old world material and immaterial aspects of culture, however, do survive wide expanses of time and space associated with this type of migration.[24]

It is difficult to put a human face on the rather complex and even theoretical topic of migration and cultural diffusion in recent and prehistoric times. Extant records and personal reflections in the form of journals are not available to give us insights into how and why places like Skara Brae, Hemudu, and Mount Sandel served as the last bastions of culture or as stepping stones of cultural diffusion.[25] Those places were not cities, but they were, by all definitions, villages. They were located at the farthest reaches of land; it was as if hunters and gathers as well as subsistence farmers had made their way to places where the land gave way to sea. To their amazement, they found the carrying capacity much higher in coastal settlements where the intermingling of fresh and salt waters formed estuaries that provided them with abundant plant and animal life. By the late Mesolithic era, improvements in technology and climatic warming, according to E. S. Deevy, enabled inland villages to support up to fifteen people per square mile, whereas the carrying capacity during the Paleolithic age was only 0.025 people per square mile.[26] The evidence found at Skara Brae, Hemudu, and Mount Sandel suggests that forested areas near coastal waters were capable of supporting a much higher population density than inland territories. Still, the lack of technology to establish agricultural practices to augment food supplies during times of scarcity pushed local populations beyond their respective areas' meager carrying capacities. Like the ancient Greeks and other, later nations, these microsocieties sent forth colonies to populate presumably less inhabited places. In each instance, the colonizers sought places in which they could carry on with their lives in a similar fashion. Their environmental perceptions were vital in deciding where to move. The migration pattern these ancient people formed was rural to rural, not rural to urban as argued by geographers like Ravenstein, Zelinsky, and Jones.

Nevertheless, new environments presented hunters and gatherers with perhaps equally new challenges because of changes in plant and animal communities. There may well have been changes in weather that required new ways of thinking and interacting with the local environmental conditions. For instance, we know that the Near East was wetter at the end of the Mesolithic era than it is in the twenty-first century, and certain places in North Africa and the Near East were even wetter and warmer in 10,000 BCE. The extent of native forest stands, absent more recent human exploitation, suggests climate change. The Dead Sea and the Sea of Galilee were certainly much deeper in the late Mesolithic age. Since the Roman occupation of Palestine, archeological remains of a once thriving fishing village—believed by many to be biblical Bethsaida—indicate that the shoreline of the Sea of Galilee has shrunk by about one mile or nearly two kilometers. (Climate change in the Near East is explored further in chapters 4 and 11.)

Immigrants and residents were obviously undaunted by these environmental changes. Although migrants and established residents may have lost contact with folks back home, they kept their knowledge and mental images with them. New information was added to the existing bodies of knowledge, and by 8000 BCE, migration was no longer the only option for villagers living under stressful conditions caused by local overpopulation. Since earlier migrations had filled nearby rural places, villagers who now found that similar niches were closed to them were forced to make adjustments in their way of life. By using knowledge of local plants and animals like wheat and barley found in the uplands of the Near East, for instance, people were soon able to produce a surplus of food that permitted some members of the growing village to pursue other activities. Specialists in the arts and crafts as well as in religion created diverse local occupations and, hence, stratified communities. It was then that villages grew into cities.[27] As they grew, rules became necessary to control both secular and religious behaviors, and so ruling classes were established. Like the other nonfarmers, rulers depended on the labor of others to survive. As modifications in social, political, and economic systems increased local carrying capacities, residents of these cities shared common traits with others over a relatively large expanse of space. These larger culture areas were identifiable civilizations.

The earliest civilizations were established in Egypt, Mesopotamia, the Indus River valley, Peru, China, and central Asia (see table 2.1). As noted earlier, it is amazing that these places, separated by thousands of miles of land and sea, came into existence within about 1,500 years of each other. Despite being the home of humans for more than ten thousand years, or since the end of the Pleistocene, no large European civilization is as old as these proto-civilizations. Perhaps it was still too cold to invite immigrants. On the other hand, lands newly exposed by retreating glaciers likely provided surplus populations from semipermanent villages like Hemdu with resources sufficient to maintain hunting and gathering lifestyles. Given the age of Mount Sandel, clustered, semipermanent settlements existed for at least two thousand years before intensive agriculture was established in western Europe.

As suggested by table 2.1, there appears to be a high spatial correlation between these places in terms of both time and distance from the Near East. Peru, however, is obviously farther away from the Near East than China or

Table 2.1 Early Civilizations[28]

Place	Date
Mesopotamia	3500 BCE
Egypt	3050 BCE
China	3000 BCE
Peru	3000 BCE
Indus River	2400 BCE
Central Asia	2000 BCE

These dates are estimates of when cultural development in these already occupied natural utopias reached a stage that the areas could be regarded by later archeologists as civilizations. I contend that the rise of civilizations was set in motion by the overpopulation of surrounding hunting territories.

central Asia, but it is an older civilization than its Asian counterparts. Indeed, debate rages over the dates of wood artifacts found at Monteverde in northern Chile and paint specks found in earthen layers believed to be seventeen thousand years old at Pedro Furada in Brazil. However, genetic evidence and ocean currents provide proof of a less controversial explanation for this situation.

As was mentioned earlier, hunters and gatherers spread out from the Near East and traversed Asian and European landmasses. They built semipermanent villages on the coasts in eastern China and western Europe at nearly the same time (ca. 7000 BCE), which interestingly enough is the approximate time when the first Peruvian villages were established. Asians and Europeans (identities not used by hunters and gatherers, but a simple means to differentiate peoples in this discussion) developed modest boat technology, but, unlike Europeans, Asians were able to follow kelp beds along the coast. Those who traveled north along the Asian coastline found the Aleutians and, over time, followed them into the Americas. This migration was aided by the Kuroshio or Japan Current that, because of the Coriolis effect, circulates in a clockwise fashion in the Pacific basin north of equatorial waters. Along the coast of North America, it is known as the California Current. This circulating current would have taken travelers into the Americas. In contrast, off-shore currents in the North Atlantic, flowing toward the east and then south along the coast, impeded Europeans from setting sail or even rowing to North America. It would have been no easy feat to use the technology available before the ninth century to attempt such a journey. Because of its extreme distance from the Near East, the Peruvian settlement is another matter—one that deserves further geographic and biological analyses.

With an understanding of ocean currents, it is not too difficult to imagine what likely took place from 13,000 to 10,000 BCE, when humans first arrived in the area today known as Peru, which was about the same time early humans arrived in Britain.[29] Under the influence of the Coriolis effect, ocean currents, as mentioned previously, circulate clockwise in the Atlantic and Pacific Oceans north of the Equator, but they circulate counterclockwise in the Southern Hemisphere.[30] This means that migrants likely rode the ocean currents that steadily increased in temperature as they approached Mesoamerica and then South America. Ecuador and northern Peru sit astride the Equator in South America, and it is in this area that the currents bend westward across the Pacific. Traveling on a short distance south along the coast, the increasingly colder ocean currents that were now flowing northwestward were of little use as a transportation medium. Perhaps hoping to find warmer waters again, they may have followed the coastline a little farther south, fighting northbound currents along the way. Instead of locating temperate, rich waters, they found increasingly colder currents flowing up from Antarctica. This flow of water is today called the Humboldt Current, and the thirsty Atacama Desert forms along the Chilean shoreline farther south. For early Asian hunters and gatherers to travel farther south, they would have had to struggle in a desert environment. Returning north to Central America would have meant fighting the force of the California Current. With ample rivers like the Maranon and

Ucayali, and with streams draining the Andean uplands, whose higher elevations also offered relief from increasingly warm temperatures in low-lying areas, the post–Ice Age area in what is today Peru must have been a natural utopia. The early rise of a civilization there would certainly suggest that its builders thought it was an ideal place to live.

This geographic explanation is supported by mitochondrial DNA identified among native populations living in Central and South America. Of the four main maternal clans to leave Asia and arrive in the Americas, one of them is not commonly found among residents of Siberia and Mongolia. In fact, this particular DNA is quite similar to Polynesians who inhabited Southeast Asia. The slight changes in the Polynesians' DNA suggest rather strongly that their later relatives once again took part in a seagoing adventure. In fact, changes in DNA suggest that the ancestors of Native Americans left Asia about thirteen thousand years ago.[31]

SUMMARY

As the older geographer E. G. Bowen pointed out, coastal settlements like Hemudu and Mount Sandel, even during the Mesolithic era, served as the last bastions of culture or as departure points for cultural diffusion. They also served as points of reentry for reverse migration flows and even for people wishing to make temporary return visits.[32] As nearby rural niches filled with people, villagers were forced to use their accumulated knowledge of local plant and animal communities as well as information gained through trade and movements of people to adopt the less attractive option of farming. This turn toward intensive agriculture and the sedentary lifestyles of cities were likely caused by overpopulation, although paleoclimatologists Gordon Childe and J. T. Meyers argue that climate change (and not the overpopulation of the area) was the primary cause of the transition from the Mesolithic age to the Neolithic era.[33] A similar argument is made by archeologist Charles Burney.[34] Indeed, Burney argues that environmental factors, such as climate change, and their slow processes deserve first place as causal agents in any study of the time period or locale in which the economy shifted from hunting and gathering to sedentary and deliberate food production. It should be pointed out that Burney's conclusion is not shared by this author. Instead, I maintain that his conclusion does not take into account the fact that, given the opportunity, hunters and gatherers who obviously create rural to rural migratory patterns shift to agriculture only when farther migration is no longer the best option for their survival. Ancient China's Hemudu and Ireland's Mount Sandel provide excellent examples of this cultural adaptation.

In chapter 3, we will look at the rise of cities. Many of the world's greatest cities have grown from small villages located in places where residents at one time lived in relative harmony with the local environment into situations in which city dwellers now rely on others to exploit the natural environment for them. The building of human settlements is occurring farther and farther away from natural utopias. As chapter 3 will show, a naturalistic lifestyle, as in the case of the hunter and gatherer, is simply not an option for most people because the best natural utopias that could serve as semipermanent fishing villages or hunting grounds are now covered by sprawling cities or otherwise used for grazing and croplands. The best places to live are already occupied.

Chapter 3

SETTLING THE
ANCIENT ECUMENE

INTRODUCTION

North America and, indeed, most of Central and South America have many undeveloped hill and mountaintop home sites that offer outstanding views of valleys, rivers, lakes, distant oceans, and open green spaces. While some of those landscapes seem inviting, they often provide few of the more important products of nature that support human life. Consider places like Mount Rogers in Virginia, Kentucky's Black Mountain, Mount McKinley in Alaska, and California's Mount Whitney: each is the highest point in its respective state, but it is also a certainty that these highland locations are uninhabited by people. Until the advent of electricity in the twentieth century, most people followed the ancients' lead in building settlements, including single-family and multifamily homes, on coastal and river plains rich in alluvium (unconsolidated sediment and eroded rock material). Nowadays, however, in many hilly and mountainous parts of the United States, rustic and stately homes are being built in heretofore rugged and uninhabited places that offer residents majestic vistas of their surroundings. There is at least one paradox in this situation: the family that chooses to live in a rustic home on a ridge top overlooking a pastoral valley probably does so because of its desire to "feel close to nature." However, for most people living in less developed countries and even in the United States, this is not practical. A mountaintop residence is even less practical for those who, for economic reasons, actually depend on nature for sustenance because mountain and ridge slopes are easily eroded of soil as water and gravity work together to make these upland places unfit for deep plow cultivation. In reality, to be able to live in a modern-ridge or mountaintop "utopia," one must have a surplus of wealth because life there depends heavily on technological innovations that can deliver abundant supplies of water and electricity as well as food and other products. The rustic nature of this lifestyle is actually an illusion. As was shown in chapter 2, Mesolithic and early Neolithic peoples like those who called Skara Brae, Mount Sandel, and Hemudu their homes knew where natural utopias, if they were to be found, were located. Such places are typically not situated in high places such as hilltops, ridges, and mountaintops. Instead, natural utopias are located in low-lying areas that offer ample supplies of water; moderate temperatures; and deep, fertile soil; as well as plants and animals to provide protein and nutrients. Ideal places are also located in the middle latitudes, near oceans with comparatively warm surface currents. Because of their low evaporation rates, cold oceanic currents like those flowing along the west coasts of continents are associated with creating arid landscapes. As this chapter will show, *Homo sapiens* have been seeking out the most beneficial climates and putting down roots in a global context for about fifty to sixty thousand years. Because the vast majority of natural utopias are already occupied, present and future generations will increasingly have to depend on innovations and technical inputs to live in higher, drier, or otherwise more environmentally challenging, even dangerous, places.

ACCOMMODATING POPULATION GROWTH

Any responsible consideration of issues like expanding food supplies, reducing pollution, and encouraging economic development often and perhaps naturally proceeds from knowledge about human population because the places where we live have limited carrying capacities. Geographers with interests in natural resource conservation, geomorphology (the study of forces that shape the landscape), and climatology recognize that asking the where and why questions of population is critical to understanding a host of human-environmental interactions and their consequences. In a global context, geographers ask "Where are people concentrated, and why are they there in such large or, alternately, scant numbers?" By knowing where human life is best supported by nature, questions about human-environmental interactions can then proceed from this foundation. Historically, humans have avoided settling in places where life is precarious or uncomfortable. Therefore, an alteration in the boundaries of the ecumene laid out by preagrarian and preindustrial people depends heavily upon innovations that inspire the imagination of possiblists and human ecologists. Climate change, too, regardless of its causation, may well force alterations in lifestyles if humans continue to live in the ancient ecumene, for many areas in the upper midlatitudes, such as those on which Aberdeen, Moscow, Stockholm, and Berlin now stand, were once covered by glaciers. There is no assurance that two-mile-thick sheets of ice will not return to occupy those densely populated places.

Looking back to the end of the Ice Age, there is some debate among scholars working in demography about the numbers of people alive at the onset or during the Agricultural Revolution and the birth of the Neolithic age. There were certainly no census officials working in bureaucratic offices or pounding the pavement, going door-to-door in 8000 BCE. However, an estimate based on reverse growth models ranges from four to five million people.[1] One way to see how important that estimate is to each of us in the present day is to consider how our own family trees relate to it. While communication technology may seem to make today's world smaller, the numbers of people living in past times, especially around the time of the Agricultural Revolution in the Near East, suggest that theirs was indeed a small world. By using an approach that maintains four generations per one hundred years, a child born in 2010 would have sixteen great-great-grandparents living in 1910. To show how small the world really is, consider that this child would have 4,194,304 ancestors living when Christopher Columbus's parents were contemplating becoming parents in early 1451. It takes only a little thought to realize that at some point in each person's family tree, relatives married or otherwise mated with kinfolk. The notion that each person is a product of a family tree that expands outward along its many branches must be challenged; at some point in the past, the tree analogy no longer accurately described familial connections, let alone the fantastic numbers of people needed to make it a realistic depiction of our collective

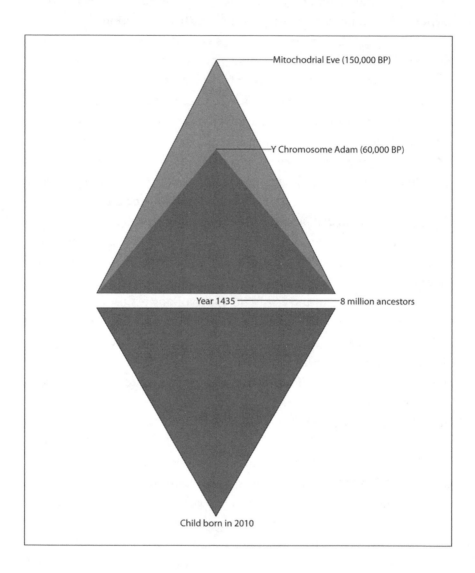

Image 3.1. Deep ancestry. While all modern humans have common parents, mitochondrial Eve and Y-chromosone Adam each lived in much different times. *Image courtesy of Geraldine Allen and Clayton Andrew Long.*

deep ancestry. In reality, deep ancestries resemble not a tree, but an inverted pyramid with an upright pyramid stacked on top of it, base-to-base. The family-tree analogy is a rather shortsighted perception of ancestral connections. It seems to work well, however, for those interested in genealogy who are able to go back up to only six or seven generations.

With the stacked and opposing pyramids in mind, the top pyramid represents each person's deep ancestry, converging on a Y chromosome "Adam" and a mitochondrial "Eve" (see image 3.1). How else could a child born in 2010 have more ancestors alive in 1435 (8,388,608) than there were people living at the time of the Agricultural Revolution only 9,500 years earlier?

Population, as Thomas Malthus observed in 1798, is growing exponentially, but such growth is not, nor has it ever been, uniform across the earth's surface. When Malthus was alive, the world population had not yet reached one billion. By 1750 and around the time Malthus was born, there were approximately 790 million people alive. By 1850 the world had reached 1.26 billion, and in 1950 the population was 2.55 billion. Today, there are more than seven billion people sharing the planet. We are adding roughly eighty million people per year.[2] It is important to know that there are more people living today than at any time in human history. Thanks in large measure to the diffusion of technology produced during the Medical Revolution that resulted from accumulated knowledge acquired during both world wars, the rate of growth was faster in the last half of the twentieth century than at any time in the past. In other words, the normal growth rates in population that have characterized demographic transitions throughout most of human history are now supercharged by medical and food-production technologies. As these technologies diffuse across the planet, oftentimes facilitated by well-meaning philanthropic organizations, population explosions occur in less developed countries, which then produce greater and more intense migration flows into already overcrowded places. Our ancestors originally chose these areas because they were natural utopias, but the numbers of people living in them now have made us dependent on others who we do not know and with whom we have little to no contact. They supply us with food, water, and clothing.

OUR ECUMENE

When the North American continent was experiencing what John L. O'Sullivan (1813–1895) called Manifest Destiny from 1845 to 1900, most people migrating west saw the Great Plains of the interior as an obstacle to reaching the West Coast and its rich natural resources in the forms of fish, whales, gold, redwoods, and Douglas firs. Ironically, prairie soils that had lain under the rolling wheels of westward-moving wagon trains included the highly fertile soils chernozems and mollisols, yet because there were few trees and little surface water (west of 100° west), pioneers regarded this section of the expanding country as poorly suited for farming. The plains remained sparsely populated as the coastal west filled with people. A look at the spatial characteristics of where humans now live yields few surprises. The fact that humans still live in places marked by their ancestors thousands of years ago is a testament to the accuracy of humans' earliest geographic perceptions and the efficacy of their migration and settlement-building decisions.

Humans are found in their highest concentrations in a relatively narrow swath of the earth's surface, roughly between 10° and 55° north latitude. Even within that modest band of earth, people are most likely to be found within three hundred miles of an ocean. A look at a world map shows that this is, in and of itself, not unexpected because North America, North Africa, and Eurasia are all located in this part of the earth's latitude. Sub-Saharan Africa, Australia, and South America are, in aggregate terms, a much smaller part of the earth's land surface. Just over two-thirds of the earth's crust is covered by water; only about 29 percent of the earth's surface is dry landmass. Despite the arguments of geographer Paul Vidal de La Blache and other proponents of possiblism, humans have not gained complete control and, hence, dominion over the forces of nature because the broad parameters of the ecumene have changed little since the Agricultural Revolution. As they were in ancient times, the highest concentrations or clusters of people are found in coastal and adjacent areas. In fact, 60 percent of modern humans live in Asia, but they are highly concentrated in the eastern and southern reaches of the continent. Meanwhile, Asia's dry interior is sparsely populated.

The earliest cities are found along water bodies in southwestern Asia. These took the form of settlements such as Damascus, Kirkuk, Jericho, and Ur and date back to about 4000 BCE—a relatively short span of time in the earth's existence.[3] As discussed in chapter 2, the construction of cities and the shift toward intensive agriculture may well have been a byproduct of choices made by people who found that migrating in a rural to rural fashion, as would be the fashion in hunting and gathering societies, was no longer their best option. There was simply too much competition for space and resources to continue living as hunters and gatherers. Although the early Peruvian civilization is a notable exception, as a general rule, younger cities are situated farther away from the Near East. This rule has led geographers and anthropologists to debate the mechanisms behind the spreading out of cities. Was it caused by diffusion as the dates seem to imply, or through multiple hearths? It seems likely that the earliest cities and their civilizations emerged independently of each other as spaces needed for those who continued hunting and gathering filled with people. The coastal and lake-shore locations of the five largest concentrations of people, both in the past and in the present, certainly put forward the idea that cities grew in response to a significant portion of people believing they had run out of places to continue life as hunters and gatherers.

Currently, the world's five largest concentrations or clusters of people are located in east Asia, south Asia, Southeast Asia, western Europe, and eastern North America, respectively. In a sense, it is as if humans, including the ancestors of Native Americans, have been queuing up for departure to some distant new land. This phenomenon is clearly seen in population densities in coastal areas in the midlatitudes. For example, in 2007, Arlington, Virginia, which is located in the Potomac tidal basin next door to Washington, DC, had a population density of 7,898 people per square mile; Boston, Massachusetts, had a density of 12,383; and Chicago, Illinois, was slightly denser with 12,491. Hialeah, Florida, had a population density of 11,503, and Miami, located on the southeastern shores of the Sunshine State, had a population density of 11,477 per square mile.[4] These cities, of course, pale in comparison to the Big Apple. In 2007, New York City had a population density of 27,282 per square mile. On the West Coast, there are

Image 3.2. Densely populated Manhattan. *Image courtesy of the author.*

Image 3.3. Atlantic City, New Jersey. *Image courtesy of the author.*

Image 3.4. San Francisco. *Image courtesy of the author.*

also densely populated urban areas. San Francisco had a density of 16,381 in 2007 while Long Beach, California, had 9,256 people per square mile.

In contrast to these coastal cities, interior cities have fewer people per square mile. In 2007, Lexington, Kentucky, had only 981; Kansas City, Missouri, 1,437; and Oklahoma City, 902 people per square mile.[5] There are interior cities with greater densities, but they are still less congested than coastal areas. In 2007, Dallas, Texas, had a density of 3,622, and Denver, Colorado, had 3,820 people per square mile.[6] Table 3.1 reveals settlement patterns in the United States from 2009. Coastal states boast an average

Table 3.1. Population Density of States by Proximity to Navigable Water Bodies

Coastal States N = 21		Great Lakes States N = 8		Inland States N = 20	
AL	91.4	IL	231.5	AZ	55.8
CA	234.7	IN	177.1	AR	54.5
CT	722.9	MI	178.2	CO	46.9
DE	443.1	MN	65.3	ID	18.1
FL	340.4	NY	194.8 (N = 25 counties)	IA	53.5
GA	166.0	OH	280.6	KS	34.0
LA	99.3	PA	277.9	KY	107.4
ME	42.7	WI	103.4	MO	85.6
MD	578.8			MT	6.6
MA	826.7			NE	23.1
MS	62.6			NV	23.4
NH	147.0			NM	16.2
NJ	1,180.8			ND	9.3
NY	578.2 (N = 37 counties)			OK	52.7
NC	186.4			SD	10.5
OR	39.0			TN	149.3
RI	1,023.3			UT	32.2
SC	146.5			VT	67.4
TX	91.5			WV	75.4
VA	195.3			WY	5.4
WA	97.3				

It is important to point out that only the forty-eight contiguous states were compared; twenty-five counties in New York, which are considered within the Great Lakes Basin, were placed in the Great Lakes category while the data for the state's remaining thirty-seven counties were placed in the Coastal States category.

population density of 347 people per square mile, while Great Lakes states have 189 people, and inland states have only 46 people per square mile.[7]

To illustrate the role of water bodies such as navigable lakes, rivers, and seas on population density, consider that when Louisville, which is situated on the Ohio River, is excluded from the land area and population data for Kentucky, the Blue Grass State's population density drops to 89.8 people per square mile (see note 7 of this chapter for a statistical analysis of the data in table 3.1). In the case of Tennessee, Hamilton, Davidson, and Shelby Counties (four of ninety-five counties in Tennessee), which are situated on the Tennessee, Cumberland, and Mississippi Rivers, respectively, these are the homes of over one-third of the state's 6.1 million residents. With respect to the state of Maine, its population is heavily clustered along the Atlantic coast. Sixty percent of Maine's population lives in the coastal areas around Augusta, Bangor, and Portland.[8]

In other parts of the world where humans have clustered for millennia, coastal areas again have much higher population densities than inland locations. Monaco, the Mediterranean microstate pressed against the sea by its northern neighbor, France, is one of the most densely populated countries in the world; the country has 43,559 people per square mile.[9] Singapore, a small, tropical country located just off the coast of the Malayan Peninsula, has some 17,482 people living in the same-size area, while the south Asian country of Maldives has 3,331 people per square mile.[10] Bangladesh, which sits precariously on the low-lying coastal plains of the Bay of Bengal, has 2,969 people per square mile. On the other hand, landlocked Mongolia, the mitochondrial home of the Cherokee Nation, had only five people per square mile in 2007. Meanwhile in that same year, Brazil, the most populous country in South America, had 60 people per square mile. Other large countries located in the Southern Hemisphere like Australia and Argentina had 7.1 and 38.3 people per square mile, respectively.[11] Even in those countries, people prefer to live in coastal settlements. For instance, Sao Paulo and Rio de Janeiro, Brazil's largest cities, are located on the shores of the South Atlantic. About 15 percent of the country's total population lives in those two cities. Brasilia, the capital of Brazil, was intentionally built in the interior of the country to reduce population pressures along the coast. In

Argentina, an astonishing 32.4 percent of its population lives in the coastal metropolis of Buenos Aires.[12]

While the automobile and systems of railroads and motorways have made interior locations accessible, helping to diffuse populations, coastal areas still attract people who are willing to cope with the associated spatial limitations. As one can see from the price of real estate in coastal areas, there is a direct relationship between price and proximity to the sea. Similar patterns can be found in cities that are situated on the shores of lakes and on bluffs overlooking rivers. Like our ancestors, it seems we still want to live in places where we can draw inspiration from sea breezes and shorelines washed by frothy surf.

Our apparent need for natural inspiration may well lie at the heart of the reasons for the lower population densities of interior settlements. Paul L. Knox and Sallie Marston rightly point out that Dallas is "a classic low-density urban settlement predicated on the widespread use of automobiles."[13] The automobile, hence, provides a means of commuting that disperses population away from central business districts (CBDs). Here, as before in human history, there are often environmental push and pull factors that influence where we live relative to the sources of our livelihood. As with ancient hunters and gatherers who seemed to enjoy home bases at Mount Sandel, Hemudu, and Skara Brae, people in the twenty-first century are perhaps willing to live in tight spaces if those spaces are located on or near large water bodies, but in the case of inland cities, they are inclined to commute longer distances from so-called bedroom communities to urban workplaces if it allows them to live a lifestyle at least partially exposed to open and seemingly natural spaces. With the advent of cars and mass transit systems, people who can afford them have moved to surrounding areas where they can be rejuvenated, enjoying green spaces and less exposure to the noises and smells associated with the artificial, anthropomorphic landscapes found in cities. The pursuit of *Gemeinschaft* in the country was perhaps the impetus for the "rural renaissance" of the 1970s.[14]

As others from the city follow, dispersed settlements around cities then become congested. Rural property owners, tempted by increasing amounts of money offered for their lands, sell them to developers who build larger communities. More often than not, single-family residential developments in the United States are given names like Brook Green, Willow Creek, Turkey Creek, River Oaks, Tara Hills, Shepherd Hills, and other names that imply a cozy country life sequestered away from environmental stresses associated with cities. Nevertheless, these rustically named places sprout up on the outskirts of larger cities. But over time, adjoining areas around them become developed, and then, as stores, offices, and schools are built to make life even easier for residents, the formerly appealing names suggesting a quiet country life eventually look out-of-place as cities grow around them. In many respects, this is a modern incarnation of the University of Chicago's Robert Parks, Ernest Burgess, and R. D. McKenzie's Concentric Zone model, which they developed to describe urban zones and land uses around Chicago in the 1920s. This model still has relevance in explaining some of the *Gemeinschaft* forces behind many urban to rural, intraregional migration decisions.[15] Clearly, this process contributes to urban sprawl.

WHERE LIFE IS HARD, FEW PEOPLE ARE FOUND

When the earliest migrants left the Near East, they came across a plethora of places where life would have been too difficult, if not impossible. As our ancestors spread across the planet, they tended to avoid putting down roots in places that were too wet, too dry, too cold, or too high in elevation. The ancients lacked the technical knowledge to make living there feasible. Even today, with our sophisticated technology, many of these places are not heavily settled, if at all. These sparsely settled places occupy a significant portion of the earth's land surface. Our ancestors' understanding and avoidance of them as places to settle demonstrates their pragmatism. If present and future generations try to live in such places, it will likely be costly, and they will perhaps suffer negative consequences. When compared to making alien worlds like Mars and Venus hospitable, however, these inhospitable areas are less costly and quite attractive. The costs are tied to the development or acquisition of technologies to reign in life-threatening forces of nature. Nonetheless, as this section shows, where life is hard, few people are found.

Places that are too wet are mainly in latitudes where low pressure centers prevail throughout the year. Humans flourish in places that have at least seasonal variations in high- and low-pressure systems. Rainy low pressure systems stretch across a swath of earth between 20° north and 20° south. They are common across the interiors of Africa, Central and South America, as well as Southeast Asia. Rainfalls in the "wet zone" average between fifty inches (1.25 meters) to more than ninety inches (2.25 meters) per year.[16] Nontechnology-dependent societies have not done well in this latitudinal zone. The Democratic Republic of the Congo in central Africa, for instance, lays claim to a meager population density of 29.6 people per square mile, but its population is denser than the nearby Central African Republic; it is the home of only 18.5 people per square mile.[17] Bolivia, located in the interior of South America, has a population density of 22.1 people per square mile.[18] Papua New Guinea, an island country off the coast of Southeast Asia, has a population density of just over 33 people per square mile.[19]

The case of Laos is interesting because it illustrates how much more attractive coastal areas are in the tropical wet zone. Unlike its northern neighbor China, however, four-fifths of Laos sits in the Northern Hemisphere's tropical wet zone. The eastern highlands, with their lush jungle vegetation, give way to the Mekong River basin in the west where the majority of people live. The population density in Laos is 74.9 people per square mile, which may seem high when compared to the Central African Republic, but in comparison to its coastal neighbors, it is sparse.[20] Laos is a landlocked nation surrounded by coastal countries. In addition to China, its neighbors are Vietnam with 685.5 people per square mile, Cambodia with 209 people per square mile, and Thailand with 331 people per square mile.[21] China, the most populous country in the world, has a regionally modest population density of 369.4.[22] The dry highlands of western China do not offer humans the natural bounty as does its eastern shoreline. The western regions of China are characterized by the Kunlun Mountains, the Plateau of Tibet, and the exceedingly dry Taklimakan Desert. The Chinese population is hence highly concentrated in the east near the Yellow, East China, and South China Seas. The population density of Laos indeed looks meager when compared to that of Hong Kong. The former

British colony and current Chinese coastal city is the home of 17,925 people per square mile.

Dry lands like the Taklimakan Desert provide settlement opportunities for even fewer people. Dry lands are created by prevailing high-pressure centers that are found on the west sides of continents between 15° and 45° north latitude and between 20° and 50° south latitude (although much of the tundra of North America and Eurasia are cold deserts—those with less than ten inches of rain per year). The Atacama Desert stretches along the west coast of South America. The deserts of Mexico, California, and Arizona are located in North America's dry region. The Sahara Desert of North Africa and the Great Australian Desert are also located in these high-pressure zones. In addition to prevailing high-pressure centers, cold, oceanic currents moving in from icy polar seas offer little atmospheric moisture to these stretches of parched land. Cold ocean currents are heat sinks, absorbing energy from both the land and the sun. Their slow-moving water molecules have little energy to force them to reach escape velocity and break free from their liquid states. The combination of prevailing high-pressure systems and cold water currents produce little humidity that can be squeezed out over scorched lands.

The vast expanse of Australia lies west of the Australian Alps. Most of the population is pressed along the east coast of the country, where rainfall is plentiful. The climate along the east coast supports lush, green forests and swampy lowlands where saltwater crocodiles serve as apex predators. On the other hand, the climate west of the Alps is quite dry. The Outback, located in the interior, is sparsely populated. Overall, Australia's population density is only 7.1 people per square mile. When the populations of Canberra, 378,000; Sydney, 4,327,000; Melbourne, 3,728,000; Brisbane, 1,860,000; Perth, 1,532,000; and Adelaide, 1,145,000 are factored out of the density figure, the number drops to 2.71. As one might expect, each of these cities is located on a coast.[23]

Chile provides another interesting example of how dry lands are not used as integral parts of the human ecumene. The hot and dusty Atacama Desert occupies much of the northern lands of this narrow coastal country. The southern part of Chile has more rainfall, so the climate there supports forest stands and grazing lands. It also supports most of the country's sixteen million people. Even with population centers like Santiago (5,720,000) and Valparaiso (854,000) included, the country's density of 56.9 people per square mile is comparatively low. If one excludes the populations of those two cities, Chile's density drops to 33.1 people per square mile.[24]

While North African countries like Morocco and Egypt are widely known for their rich contributions to Western civilization, most of their populations are clustered in cities along the Mediterranean Sea or on the banks of the region's few rivers. Since the early 1300s, Cairo, located along the Nile, has served as an important city for the desert nation. From a population of roughly 4,000 in the fourteenth century, Cairo has grown to more than fifteen million, making up some 20 percent of the country's population.[25] Alexandria, the site of the finest library in the ancient world, is located just north of Cairo, at the mouth of the Nile. The Mediterranean city is the home of another 4.1 million Egyptians.[26] Incidentally, Cairo is one of the most densely populated settlements on earth. Whereas New York City has a population density of 27,282 per square mile, the Egyptian capital is the home of 83,000 people per square mile.[27] The city continues to grow, gobbling up some of the most fertile lands in North Africa. As Malthus predicted, the growth in population is outpacing the city's ability to employ the current population, let alone newcomers.

Other less densely populated areas have developed across North Africa in oases that form in low, wet areas in the Sahara Desert. There are even more sparsely populated countries in Africa, but they are in the Southern Hemisphere, away from the historic ebbs and flows of Western civilization. While Egypt is located in the Northern Hemisphere with population centers sitting at latitudes similar to Austin, Texas, and Mobile, Alabama, Botswana and Namibia are bisected by the Tropic of Capricorn (23.5° south) in the Southern Hemisphere. The Kalahari and Namib Deserts form much of these two nations' landscapes, and they have meager population densities of 8.2 and 6.6 people per square mile, respectively.[28]

Cold places are also less settled than the temperate environs of mid-latitude, coastal areas, including important estuaries like the Chesapeake Bay and the Hudson River. Despite thousands of miles of shoreline, Alaska has fewer than two people per square mile. The rich and varied landscape of Montana is the home of only 6.6 people per square mile. These numbers pale in comparison to even the most rural of the states east of the Mississippi River. Located along the banks of the interior-draining river, Mississippi has fewer people per square mile than any of the eastern states, except for the much colder state of Maine. Rural Mississippi has a density of 62.2 people per square mile. Located in the extreme northeastern corner of the United States along the rugged Atlantic coast, Maine has a modest population density of 42.7 people per square mile.[29] Given that most of Maine's population is concentrated on or near its eastern shoreline, the hilly and more rugged interior of the state is sparsely inhabited.

With most of its population clustered within two hundred miles of its border with the United States, especially in the eastern Great Lakes area and along the Pacific coast, Canada has only 9.5 people per square mile. Similarly, the North Atlantic island nation of Iceland has only 7.9 people per square mile.[30] These numbers have much more meaning when one compares them to India, which has a density of 1,000; or to South Korea with its 1,276.1 people per square mile.[31] As peninsulas, India and South Korea, both located in the Northern Hemisphere between 10° north and 40° north, have for millennia provided humans with natural utopias in which to settle.

Humans have also historically avoided living in highlands. Many geography students are amazed to learn that there are alpine glaciers located within a dozen or so degrees north or south of the Equator. Mountains like Tanzania's Kilimanjaro (19,340-foot elevation) and Ecuador's Chimborazo (20,702-foot elevation) are crowned by impacted ice and, therefore, offer little beyond an excellent view to people in search of a permanent home. The Himalayas in Asia, the Andes in South America, and the mountains of the northwestern portion of North America have likewise attracted few people. On the other hand, when elevation and latitude work together to create comfortable living conditions, people are keen to settle down in such places. Few large cities, however, are located at high elevations in the midlatitudes and upper latitudes (30° to 90°). Situated just south of the Tropic of Cancer, Mexico City sits at an elevation of 7,200 feet. With nineteen million people, it is the world's third most populated city. This elevation, however, would not attract as many people if the city were located in the upper midlatitudes. Only a few highland areas in the United States have settlements of at least one hundred thousand people.

Denver and Colorado Springs, Colorado, are two such places. They are located just east of the Front Range of the Rockies on the High Plains at an elevation of 5,269 and 6,200 feet, respectively. Denver, the Mile High City, has a population of 588,000, and Colorado Springs has over 350,000 people.[32] Their sites and situation features are unique and make them much more comfortable than their elevations and latitudinal locations would suggest. The land is fairly flat around them, but the Front Range provides something of a windbreak from the sometimes fierce westerly winds that howl across the open plains of eastern Colorado, western Kansas, and western Nebraska. In fact, the average wind velocity in Denver is 8.7 mph. To put that wind speed into perspective, consider that it is similar to the wind velocities that gently blow across Portland, Maine; Pittsburgh, Pennsylvania; and Nashville, Tennessee. In contrast, Chicago has an average wind velocity of 10.3 mph. Interestingly, Chicago's average wind speed is slower than the air currents that blow across Buffalo, New York (11.8); Wichita, Kansas (12.2); and Cheyenne, Wyoming (12.9).[33] Under certain barometric conditions, the Front Range also enables dry mountain air to warm as it sinks down its eastern slope. Even in the darkest days of winter, these atmospheric conditions offer residents of Denver and Colorado Springs a welcome respite from the harsh climate typically found at comparable elevations in the midlatitudes, including those located only two hundred miles east of them on the High Plains.

These unusually large cities do not share some of the other limiting factors typically associated with highland locations. As a general rule, highlands exist because the underlying rock is hard and resists erosion. Aside from flat to gently rolling tablelands associated with plateaus, most highlands are steep and not suitable for travel or deep-plow cultivation. More often, soils are shallow and stream velocities are high, so along with sometimes frigid weather conditions in winter months, most highlands are sparsely inhabited. Denver and Colorado Springs are, without a doubt, highly unique.

Image 3.5. An aerial view of the Front Range of the Rockies. *Image courtesy of the author.*

Image 3.6. Northern California's costal range. *Image courtesy of the author.*

Colorado's northern neighbor Wyoming is more typical of highland areas in the midlatitudes. With only 522,830 people living on its 97,100 square miles, it has an even lower population density than Montana, as well as a lower total population than Denver. Laramie is the state's most populous county with a population of 85,296. While the state's population density is 5.4 people per square mile, Laramie County's figure of 31.76 does not conjure up images of a sprawling urban landscape even though Cheyenne, the state capital, is situated within it.[34]

SUMMARY

As this chapter shows, humans live in places where life is most comfortable. The earliest migrants who left the Near East about fifty thousand years ago were astute in their ability to find places where later generations put down roots. These were spaces where life could be lived at its fullest. More often than not, and despite the Agricultural and Industrial Revolutions, people have continued to live near oceans, lakes, and rivers in the northern low latitudes to midlatitudes. Over the years, these same places have served as stepping stones of cultural diffusion or as the last outposts of civilization. With accumulating knowledge of local flora and fauna and population pressures resulting from seemingly few opportunities to continue with hunting and gathering lifestyles, cities and intensive agriculture grew in places where people were already inclined to stay. As they have throughout history, people do not live well in places that are too wet, too dry, too cold, or too high. However, as populations continue to increase and with few natural utopias left unoccupied, societies around the world will necessarily have to adapt to life in less desirable places. To our advantage, these places are already within the broad parameters of our historic ecumene.

Chapter 4

GEOTHEOLOGY FROM THE CRADLE OF CIVILIZATION

INTRODUCTION

The Near East or the heart of the Middle East is the hearth or point of origin for all peoples. It is the earliest location of human civilization. It is also the birthplace of Judaic monotheism and its two universalizing progeny: Christianity and Islam, the world's largest and most geographically expansive religions. Long before Abraham (ca. 2000–1900 BCE), the patriarch of those religions, trod the banks of the Euphrates in search of a Promised Land, Mesopotamian civilizations flourished in the eastern Fertile Crescent while the western reaches of the Crescent served as the safest passage into and out of Egypt and the rest of Africa. The Sumerian society in which Abraham settled near the confluence of the Euphrates and Tigris Rivers featured a host of nature gods. Among them were Ki, goddess of the earth; Nammu, goddess of the sea; An, god of heaven; Nanna-Sin, god of the moon and patron of Ur, one of the world's oldest cities and Abraham's adopted home; and Enlil, the god of winds and storms. From a purely functionalist perspective, the established polytheistic religion, like the various forms of monotheism that followed it, offered adherents a psychological framework through which they could cope with and understand nature and its forces. Polytheistic and monotheistic religions in ancient and medieval societies of the Near East embraced the belief that deities controlled winds, rain, dry spells, and even tectonic activity and often used them to punish or reward segments of humanity.

While Abrahamic religions arguably ushered in an era in which most Near Eastern people jettisoned nature gods in favor of a personal deity who transcended all realms of life, control of natural forces was nevertheless retained by the God of Abraham. Clearly the notion that any god could or would use the forces of nature to punish or reward people, including nations, is a vapid assertion according to people with more secular minds. Nevertheless, as this chapter will show, texts of the Abrahamic religions, which were written over the course of two millennia, offer evidence that ancient and medieval people had a solid appreciation or fear of many aspects of the natural processes they experienced. They also provide tantalizing glimpses into climatic conditions at the time each respective author penned his book. The sacred texts of the Abrahamic religions hence tell us much about how the ancient hearth of humanity understood and thus coped with the forces of nature.

Even highly secular people of today adhere to beliefs that nature punishes people for the "sins" of thoughtless, greedy exploitation, and excess, albeit absent the supernatural attributes of one or more deities. The most obvious example of this type of nature-punishes-human scenario is seen in the climate change debate. With or without the Mesopotamian nature gods or the God of Abraham, the belief that nature punishes or rewards people for their behaviors toward it seems to be a resilient trait found among religious and secular peoples.

As a discipline, geographers have more often than not concerned themselves with studying problems that lend themselves to mapping and spatial quantification of natural or cultural features. Because geographers map and measure observable and existing features, studies are most often set in a modern context. Until recently, geographers have avoided delving into the thought processes of religious people perhaps because they have regarded such activity as falling within the area of study relegated to philosophers and theologians. As the next section will show, the existence of texts and archeological evidence from ancient times has enabled geographers to develop a subfield that considers the interfaces between the imagined worlds of religionists and the ways in which adherents have understood human-environmental interactions. Still, few geographers attempt to enter the minds of long-dead people whose actions were influenced by deeply held beliefs. This chapter fills in some of the void in those neglected and important aspects of human-environmental interactions, for many of those beliefs are still with us.

GEOTHEOLOGY

Concepts drawn from the emerging specialty area that geographer Roger Stump calls religious geography provides the framework for what lies ahead in this chapter.[1] The immaterial aspects of religion—namely, imagined worlds infused with geopious imagings (thoughts that tie together the emotional side of religion to spaces)—can influence all manner of human behaviors that have the potential to impact the landscape and our relationship to it. Particularly important in helping to frame what follows are geographers of older vintage, such as Carl Sauer, E. G. Bowen, and John K. Wright. "The geographer," Sauer argues, "is engaged in charting the distribution over the earth of the arts and artifacts of man, to learn whence they came and how they spread, what their contexts are in cultural and physical environments."[2]

To delve into Sauer's understanding of cultural diffusion, which is a process we have and will continue to witness in the spreading out of human populations from the Near East, it is helpful to enter the religious-environmental thought worlds of those who have diffused culture. Furthermore, inspiration and perspective on this important task can be drawn from E. G. Bowen, a Welsh historical geographer who, it has been argued, was influenced by Sauer and appreciative of the historical and cultural perspectives in Sauer's famous 1941 paper, "Forward to Historical Geography."[3] Bowen held a strongly cultural, including religious, conception of spaces, which he displayed in his presidential address to the Institute of British Geographers in 1959. Therein he described Wales as a highly unstable spatial construct varying in response to all manner of cultural influences coming and going across the centuries.[4] According to Bowen, both Celtic pagans and Christian religionists have shaped Wales's religious landscape over the centuries. Like Sauer and Wright, Bowen's ideas have relevance for the study of other religious expressions; but outside of a few scholars, most of their ideas have been applied only to Christianity and Islam.[5] Still, as important as these geographers have been in forging a new geography of religion, their approach falls short of truly entering the minds of those who actually shape the landscape and, along the way, alter or create both secular and sacred spaces. In this area of inquiry much has been accomplished by the likes of geographers John K. Wright, Yi-Fu Tuan, me, and historians such as Avihu Zakai.[6]

In this chapter, several geotheological terms are used that were coined by John K. Wright. Because those terms and concepts have relevance to religious geography, the research reported in this chapter owes much to the distinctive and neglected synthesis of Wright's geosophy and R. W. Stump's description of religious geography.[7] Specifically, Wright's taxonomy includes geotheology (the general relationship between space and the worship of God), geopiety (religious, emotional attachment to space), geoteleology (the unfolding of Providence related to space), and geoeschatology (the role of space in the outcomes of Providence), which are all significant in what follows.[8] Wright's approach and his useful taxonomy call upon us to consider such interior imagings in geographical research because of the role that imagings, feelings, and thoughts associated with religion have in molding a host of human behaviors that often impact the landscape with visible patterns and structures.[9] This chapter lightly embraces aspects of geographers Carl Sauer and Wilbur Zelinsky's spatial delineation of religion, but then moves on to a position more closely akin to Wright as a way of deepening the explanation that is offered for the religions' understandings of nature and its relationship to humanity.

Although geotheology is well suited to the study of the interplay between religious thought worlds and space, including nations, its conception thus far in the discipline has not offered sufficient subconcepts to capture the more subtle belief that nature can be used by divine forces to punish or reward humans—though I made a rudimentary attempt to do so in my study of Scottish, Irish, and American geotheologies in the seventeenth and eighteenth centuries.[10] The discipline needed a couple of new terms to frame the notion that God uses nature to reward or to punish people. In the spirit of John K. Wright, this situation can be remedied rather easily. The terms *geotheokolasis* and *geotheomisthosis* fill that void in the geographic lexicon. Geotheokolasis unites the Greek concepts for earth, god, and punishment while geotheomisthosis also uses Greek words to conjoin earth, god, and reward or payment. To account for more secular minds, the terms are modified to geokolasis and geomisthosis to describe the increasingly popular view that nature, acting without god, punishes or rewards people. Before we delve into those topics, however, it is helpful to see how the Abrahamic religions regarded the earth and the forces of nature. Because the sacred texts of those three religions were written over an approximately two-thousand-year span of time, the way in which weather, climate, and tectonic forces are used by the various authors tell us much about how they saw natural conditions in the Near East. As is seen in the last section of this chapter, it is revealing to find that certain climate conditions are seen in early texts but not in later writings. These images, I believe, suggest regional climate change.

COSMOLOGY AND GEOESCHATOLOGY

Most religions feature a set of beliefs that explains the origins or cosmology of the earth and its occupants. Many, especially Abrahamic religions, also delve into the imagined parameters of eschatology involving the end of days on earth and the beginning of human life in heaven or hell. Why are these aspects of religion important? The answer rests with the argument that there is no better place to start to understand how ancient peoples coped with nature than at the original condition of created space. It is in that state that we find our ancestors' closest connection to the creator, and, more importantly, such an image provides the basis for the reconstruction of sacred space over time. To Abrahamic fundamentalists, any alterations in religiously inspired images of space must be the result of punishments or rewards given to humanity or segments of it. As the highest form of creation, Abrahamic religions see humanity as inextricably tied to a benefits-reward scenario. All other creatures are unable to elicit God's wrath or reward for moral or immoral thoughts and public behaviors. Therefore, any alterations from the pristine condition of original creation are the result of judgments of God on wayward people, towns, and nations; hence, these religions in their purest forms insist on both personal and communal adherence and submission to God's will.

In Judaism, Christianity, and Islam, the God of Abraham is the creator of the heavens and earth.[11] Genesis 1:1 tells us: "In the beginning God created the heaven and the earth." While secular scholars scoff at any alleged accuracy of biblical passages with respect to scientific knowledge of the earth's past, the Genesis account of creation nevertheless parallels what we understand about the introduction and establishment of inanimate materials and life forms. While religionists and scientists differ on the timeline of those events as well as the catalyst for them, the Genesis account very generally follows what ecologists know about the establishment of ecosystems on the planet, including the late arrival of humans. We are, in both scientific and Abrahamic terms, earthly newcomers. What is even more amazing is the fact that the only recognizable location for the rivers that flowed through Eden to water the garden of Adam and Eve is Ethiopia. As found in Genesis 2:13, "And the name of the second river is Gihon: the same is it that compasseth the whole land of Ethiopia." The watershed in Ethiopia mostly drains into the Rift Valley Lakes. According to Spencer Wells, the noted geneticist working in paleoanthropology, the mitochondrial and Y chromosome DNA origins of modern humans is in that well-watered place.[12] Given the size of the planet and absent any supernatural explanation, humans in the time of Moses, the most likely author of the book of Genesis, would have had to retain some tradition or cultural memory, passed from generation to generation, that their roots were in Ethiopia; however, many biblical scholars use the headwaters of the Tigris and Euphrates as their guide to locate the Genesis garden. Consequently, they often place it in either southeastern Turkey or northwestern Iraq. The Genesis account, it could be argued, is simply describing the parameters of the original ecumene that stretched from Ethiopia through what we call the Fertile Crescent, the cradle of civilization.

The book of Genesis describes the earth as existing without form, a void before God acts to bring forth life on the planet. However, by the end of the sixth chapter in Genesis, God decides to destroy humans and creatures via the Noachian flood, for the wickedness of humans had made God sorry he made them.[13] The flood event as described in Genesis and in the Qur'an is not, however, the ultimate fate of the earth and its dwellers.

The book of Daniel reveals an end-time scenario in which a divinely appointed ruler will reign over the earth and its inhabitants. Daniel revealed his geoeschatological vision: "And there was given him dominion, and glory, and a kingdom, that all people, nations, and languages, should serve him: his dominion is an everlasting dominion, which shall not pass away, and his kingdom that which shall not be destroyed."[14] This geoeschatological passage is quite similar to one found in the Christian book of Revelation: "And the seventh angel sounded; and there were great voices in heaven, saying, the kingdoms of this world have become *the kingdoms* of our Lord, and of His Christ; and He shall reign forever and ever."[15]

Like Judaism and Christianity, Islam looks forward to a day in which human residents of the world will be judged according to each individual's works. That day is Judgment Day; thus, humanity's earthly existence has a beginning and an end. Muhammad provided a clear description of the joining together of those two distinct end points in earthly time. "And He it is Who has created the heavens and the earth with truth; and on the day He says: Be, it is. His word is the truth, and His is the kingdom on the day when the trumpet shall be blown; the knower of the unseen and the seen; and He is the Wise, the Aware."[16] Muhammad wrote: "All praise is due to Allah, Who created the heavens and the earth and made the darkness and the light. . . . He it is Who created you from clay, then He decreed a term; and there is a term named with Him; still you doubt. And He is Allah in the heavens and in the earth; He knows your secret [thoughts] and your open [words], and He knows what you earn."[17] As we will see shortly, the idea of earning rewards or punishments has relevance to our discussion on God's use of nature and it forces.

To Muhammad and Old Testament figures like David and Jeremiah, all secular objects, including natural features of lands and the seas as well as human events, are governed by God, so when the end-time arrives, his judgments will be just and irreversible. This tenet is consistent with Christianity and Judaism.[18] "And to Him belongs whatever dwells in the night and the day; and He is Hearing, the Knowing."[19] Indeed, according to Muhammad, "Allah's is the kingdom of the heavens and the earth and what is in them; and He has power over all things."[20] In Surah 16, Muhammad told that Allah is responsible for our enjoyment of nature's bounty. "He it is Who sends down the water from the cloud for you; it gives drink, and by it (grow) the trees upon which you pasture. He causes to grow thereby herbage, and the olives and the palm trees, and the grapes, and of all the fruits; most surely there is a sign in this for a people who reflect."[21] Also, every living thing in the heavens and the earth makes obeisance to Allah alone, including the angels.[22] Muhammad warned his followers that they must "guard . . . against a day in which you shall be returned to Allah; then every soul shall be paid back in full what it has earned, and they shall not be dealt with unjustly."[23] In Judaic verses penned by King David, a similar role of God as judge is revealed. As David wrote, "Therefore the ungodly shall not stand in the judgment, nor sinners in the congregation of the righteous. For the Lord knoweth the way of the righteous: but the way of the ungodly shall perish."[24] In the book of Proverbs, the author (most likely Solomon) painted a similar vision of a judging God. "Many seek the ruler's favor; but every man's judgment cometh from the Lord."[25]

In many respects, the geoeschatology of Islam sounds like that of Christianity. Whereas paradise is a place of reward for those who believe and submit to the will of Allah, fire is the punishment of those who disbelieve.[26] Muhammad used metaphors involving environmental features to show how a person can slip into hell. As he wrote, "Is he, therefore, better who lays his foundation on fear of Allah and (His) good pleasure, or he who lays his foundation on the edge of a cracking hollowed bank, so it broke down with him into the fires of hell; and Allah does not guide the unjust people."[27] In the Gospel according to Luke in the Bible, chapter 17, verse 27, Luke retold the story of the flood event and how the people were preoccupied with worldly matters, meanwhile Noah entered the ark and found salvation from the rising and relentless waters sent by God. In the same chapter of the Gospel according to Luke, the author reminded his readers what had befallen the people of Sodom for their inequities. The people of Sodom and their dwellings were destroyed by fire and brimstone from heaven, which sounds a great deal like a pyroclastic surge emitted during a volcanic eruption.

While destruction was Sodom's earthly eschatology, the last book in the Bible describes a global geoeschatology. John the Apostle wrote the book while in exile on the Isle of Patmos, located in the Aegean Sea. It was likely written in the last decade of the first century. In chapter 8 of Revelation, John described an end-time event in which a great mountain burns with fire. Apparently there is an explosion because the fire is cast into the sea where it kills one-third of all aquatic life along with destroying one-third of the ships. Again, this event resembles a volcanic eruption with a significant amount of pyroclastic activity. As horrible as this depiction of the future may be, it pales in comparison to that which is described later in the chapter, in verse 18:

> And the nations were angry, and thy wrath has come, and the time of the dead, that they should be judged, and that thou shouldest give reward unto thy servants the prophets, and to the saints, and them that fear thy name, small and great: and shouldest destroy them which destroyed the earth. And the temple of God was opened in heaven, and there was seen in his temple the ark of his testament: and there were lightnings, and voices, and thunderings, and an earthquake, and great hail.[28]

Muhammad described those who are condemned to hell by writing the following: "And certainly We have created for hell many of the jinn [or genies] and the men; they have hearts with which they do not understand, and they have eyes with which they do not see, and they have ears with which they do not hear; they are as cattle, nay, they are in worse errors; these are the heedless ones."[29] To Muhammad and John the Apostle, images of hell are enough to instill fear in the hearts of those who believe, galvanizing them to a life bent on service to the creator and owner of all things.

The Qur'an goes beyond telling about the fires of hell and describes some of the physical features of the final destination of unbelievers. As Muhammad warned, "And surely Hell is the promised place of them all [who disbelieve]; It has seven gates; for every gate there shall be a separate party of them. Surely those who guard (against evil) shall be in the midst of gardens and fountains; enter them in peace and secure."[30] It is clear that the hereafter presents two distinct spaces for those who have died. Unlike the Christian Bible, the Qur'an elaborates more on the polity of the final abode of unbelievers. It seems there are nineteen angels overseeing it as "wardens of the fire."[31] Unbelievers will be cast into hell where none will be spared and no aught overlooked, for the fire scorches the mortal.[32]

Despite burning images of hell, much is said in the Muslim holy book about gardens of paradise. When there is a mention of these pristine places, though, Muhammad quickly moved on to a more frightful geoeschatology laid out for unbelievers.[33] According to Muhammad, only Allah, through his gracious will, decides who can enter the garden of paradise. Muhammad pointed out, "And wherefore did you not say when you entered your garden: it is as Allah has pleased, there is no power, save in Allah?"[34] Muhammad clearly saw agricultural production and, as we will see in a later section in this chapter, destructive fires caused by lightning as evidences of the power of Allah to save and to destroy.

The Christian Bible also describes a frightful geoeschatology in which God judges humans with consequences for eternity. As is written in the book of Revelation,

> I saw a great white throne, and him that sat on it, from whose face the earth and the heaven fled away; and there was found no place for them. And I saw the dead, small and great, stand before God; and the books were opened: and another book was opened, which is the book of life: and the dead were judged out of those things which were written in the books, according to their works. And the sea gave up the dead which were in it; and death and hell delivered up the dead which were in them: and they were judged every man according to their works. And death and hell were cast into the lake of fire. . . . And whosoever was not found written in the book of life was cast into the lake of fire.[35]

These biblical and Qur'anic passages deviate little from each other in their descriptions of God's use of natural elements in end-time scenarios. In all cases, natural forces are clearly used to punish wayward humans. In an ironic twist, and to the delight of believing environmental determinists, Revelation 11:19 declares that God will punish people for destroying the earth.

As discussed in chapter 1, much is owed to the geotheological lexicon created by John K. Wright. His taxonomy helps frame the religious perspectives on the relationship between space and end-time scenarios, but geoeschatology and geoteleology (the unfolding of Providence related to space) lack application to more temporal uses of natural forces by God to punish or reward human behaviors. This understanding calls to mind the stimulus-response features of the operant conditioning model proposed by psychologist B. F. Skinner. Learning theorists call Skinner's theory *behaviorism*. In 1996, I extended elements of Skinner's model to describe how an individual's reflections and physiological reactions to his or her thoughts (self-rewarding or self-punishing) also have the capacity to modify behaviors without external stimuli. This notion is handily placed in the category of cognitive-behaviorism. In reflecting on how Abrahamic texts use natural forces to reward or punish humans for earthly conduct, it is helpful to recognize that mental images of what could happen via floods, drought, earthquakes, and fire for sinful behaviors are enough to create anxiety in some people. It follows that to reduce or avoid that anxiety, a person could arguably avoid the sinful behaviors that invite God's use of natural forces to punish the self as well the individual's community, town, or nation. Considering Wright's useful but limited conceptualization of geotheology and the theoretical construct of behaviorism, it seems that new terms are needed to describe how adherents in the Abrahamic religions, or any others for that matter, perceive the synergy of natural and divine forces aimed at correcting or reinforcing behaviors before Judgment Day. It is to that discussion that we now turn our attention.

GEOTHEOKOLASIS AND GEOTHEOMISTHOSIS

In Genesis 1 and Surah 2, Moses and Muhammad described the creation of Adam's garden. In his original state, Adam was healthy and obedient to Allah. His pristine garden provided for all of his needs, but in verses 35 and 36 of Surah 2, Allah is seen giving Adam directions on what and, indeed, what not to eat. This account is similar to that which is found in Genesis 2:17. These directions were clearly not followed, for the Qur'an swiftly declares that Adam and his unnamed kin were immediately banished from the bounty of Allah's garden.[36] The Qur'an's retelling of the Genesis account of creation and Adam's original sin is rather brief. As with the Genesis account of humanity's early existence, our flawed condition resulting from selfish choices and actions has plagued us ever since; this episode in Abrahamic cosmology established the beliefs in geotheokolasis and geotheomisthosis, the rewards-punishment plan used by God. In the book of Leviticus, we find Moses explaining God's method of rewarding people for their obedience, and in doing so, he rather ingeniously shows a method promoted by the US Department of Agriculture to ensure that soil can be rejuvenated and, thus, be used in a sustainable way to support life:

> Wherefore ye shall do my statutes, and keep my judgments, and do them; and ye shall dwell in the land in safety. And the land shall yield her fruit, and ye shall eat your fill, and dwell therein in safety. And ye shall say, What shall we eat in the seventh year? Behold, we shall not sow, nor gather our increase. Then I will command my blessing upon you in the sixth year, and it shall bring forth fruit for three years. And ye shall sow the eighth year, and yet of old fruit until the ninth year; until her fruits come in ye shall eat of the old store.[37]

Moses clearly recognized the importance of letting land rest in the seventh year so that it could rebuild its supply of nutrients. Unfortunately for the farmers and ranchers tilling the soil in the Great Plains during the 1930's Dust Bowl, they did not heed his advice.

In Islam, there is no way to be freed from the consequences of sin except through Allah's good pleasure and his free will, which depends heavily on earning his favor. To say that the Torah, the Christian Bible, and the Qur'an use fire and water as the basis for kolasis and misthosis is not an understatement. Perhaps nowhere in the Muslim holy book is this more evident than in Surah 9: "Allah has promised to the believing men and believing women gardens, beneath which rivers flow, to abide in them, and goodly dwellings in gardens of perpetual abode; and best of all is Allah's goodly pleasure—that is the grand achievement."[38] Not content with resting on the blessed promise, Surah 9 tempers the blissful image of a future paradise for those deserving of Allah's good will and grace by issuing a command to Muslim leaders: "O Prophet! Strive hard against the unbelievers and the hypocrites and be unyielding to them; and their abode is hell, and evil is the destination."[39] It is worth pointing out that Muhammad's method of evangelism is made obvious in this passage. This chapter in the Qur'an also suggests that true believers are not to worry about the plight of unrepentant souls.[40] When calamities like hurricanes strike the non-Muslim world, puritans will not likely be inclined to provide relief to recipients of Allah's geotheokolasis. To the puritan Muslim, there cannot be an innocent victim of nature.

The book of Lamentations in the Bible also shows how God uses weather to punish wayward people, including nations. The book was most likely penned by Jeremiah around 586 BCE. After witnessing the fall of Jerusalem, he clearly believed that God was punishing the nation of Israel. Not only was the nation's entire social world turned upside down with former servants ruling over the people of Israel.[41] but there must have also been a protracted drought accompanied by extreme heat. As observed in verse 10, "Our skin is black like an oven, because of the terrible famine." Jeremiah ended his book with this conclusion: "Turn thou us unto thee, O Lord, and we shall renew our days as of old. But thou has utterly rejected us; thou art very wroth against us."[42]

In a manner that is consistent with cognitive-behaviorism, both the Bible and the Qur'an ask adherents to look at their hearts (or minds) and know that God also punishes stubbornness and rebelliousness. In Psalms chapter 78, David revisited the ways in which God had used nature to punish and reward the children of Israel. In bringing Moses and his followers out of Egypt, David recalled that:

> Marvelous things did he in the sight of their fathers, in the land of
> Egypt, in the field of Zoan. He divided the sea, and caused them to
> pass through; and made the waters to stand as a heap. In the daytime
> also he led them with a cloud, and all the night with a light of fire. He
> clave the rocks in the wilderness, and gave them drink as out of the
> great depths.[43]

In a clear case of geotheokolasis, David used contrasting geotheological imagery to tell the Israelites, whose disobedience he feared was inviting God's wrath, to recall how God had used the forces of nature to coerce the leaders of Egypt to let the children of Israel leave Africa and return to the land promised to them through God's benevolence to Abraham and Jacob.[44] In Exodus 9:18–19, Moses declared that God will send a hailstorm on the Egyptians: "Send therefore now, and gather thy cattle, and all that thou hast in the field; for upon every man and beast which shall be found in the field, and shall not be brought home, the hail shall come down upon them, and they shall die." In Exodus 9:25–26 Moses described how God punished the Egyptians while sparing the children of Israel from the hail. "And the hail smote throughout all the land of Egypt all that was in the field, both man and beast; and the hail smote every herb of the field, and brake every tree of the field. Only in the land of Goshen, where the children of Israel were, was there no hail."

As in the Judaic scriptures, God in Islam is seen as didactic and quite willing to use the environment or forces of nature to punish evil men, or, on the other hand, provide them with gardens under which flow life-giving waters. For a dweller of a desert land, as Muhammad clearly was, lush gardens and flowing water must have been seen as the basis for the good life. However, Muhammad, as a believer in the importance of the Mosaic laws, was not alone in seeing decrees of divine justice in the natural world. For many years in the United States, some Christians have believed that storms and other destructive events in nature were regarded as "acts of God." The phrase has become part of the legal lexicon of American jurisprudence.[45] It may be one of the few existing concepts associated with secular life in America that the puritan Islamic community will recognize and support.

Nevertheless, to the puritan Muslim and his *umma* there can be little doubt that both the good and the bad that come by way of nature are known and approved by an omnipotent Allah. "And with Him are the keys of the unseen treasures—none knows them but He; and He knows what is in the land and the sea; and there falls not a leaf but He knows it, nor a grain in the darkness of the earth, nor anything green nor dry but (it is all) in a clear book."[46] In writing to his followers, Muhammad warned them, "Do you not see that Allah created the heavens and the earth with truth? If He please He will take you off and bring a new creation, and this is not difficult for Allah."[47] Allah is able and willing to change his mind about dispensing blessings to his believers in their earthly existence. Muhammad even reminded himself saying, "Maybe my Lord will give me what is better than your garden, and send on it a thunderbolt from heaven so that it shall become even ground without plant. Or its waters should sink down into the ground so that you are unable to find it."[48]

Qur'anic images of geotheomisthosis and geotheokolasis often bring to mind gardens, fires, and even the dark of night. Each is used to contrast the fates of both the good and the evil:

> For those who do good is good (reward) and more (than this); and blackness or ignominy shall not cover their faces; these are the dwellers of the garden; in it they shall abide. And (as for) those who have earned evil, the punishment of an evil is the like of it, and abasement shall come upon them—they shall have none to protect them from Allah—as if their faces had been covered with slices of dense darkness of night; these are the inmates of the fire; in it they shall abide.[49]

While those depictions are vivid illustrations of geotheomisthosis and geotheokolasis, there are perhaps few better examples of divine punishment on a sinful people by natural means than the story of the great flood. In some of these images, one can see strong implications for theocratic control of society. Noah (in Arabic, *Nuh*), as the book of Genesis declares, built an ark to save himself and his family from a watery death. According to the Qur'an, Nuh warned his fellow neighbors but they refused to listen to him and submit themselves to Allah. "But they rejected him," wrote Muhammad, "So We delivered him and those with him in the ark, and We made them rulers and drowned those who rejected Our communications; see then what was the end of the (people) warned."[50] Muhammad believed that others survived the flood event and became the people ruled by Nuh and his family who had survived on the ark. Apparently, survivors were not given Nuh's warning to submit to Allah, less they suffer the consequences of failing to surrender to Allah. The geotheomisthosis scenario portrayed in the flood event portends a political vision of the future, for those who have submitted to Allah are to be the rulers over others. For now, however, let us take a closer look at divine wrath and blessings administered through natural forces.

As the product of a dweller in a desert environment, it is not too surprising that the Qur'an also uses water in images of geotheomisthosis. Images depict life-giving water coming from clouds and flowing through gardens. "Who made the earth a resting place for you and the heaven a canopy and [Who] sends down rain from the cloud, then brings forth with it subsistence for you of the fruits; therefore do not set up rivals to Allah while you know."[51] Muhammad clearly saw flowing rivers and gardens rising on their shores as rewards for carrying out the duty of a Muslim. "But as to those who are careful of (their duty to) their Lord, they shall have gardens beneath which rivers flow, abiding in them; an entertainment from their Lord, and that which is with Allah is best for the righteous."[52]

Not only was Muhammad convinced that water from the sky is a blessing from Allah, he believed that other celestial bodies were designed for human use, although most assuredly under the auspices of Allah. "And He has made subservient for you the night and the day and the sun and the moon, and the stars are made subservient by His command; most surely there are signs in this for a people who wonder."[53] The seas are made to provide believers with "fresh flesh" and ornaments from the depths.[54]

Muhammad also recognized the role played by stars in helping ships to cleave the sea.[55] The Qur'an further points out that Allah's world is created with various hues, so it is a pleasure for the eye to behold.[56] In all of the bounty found in Allah's world, Muhammad insisted that the creator's divine hand can be seen. This insight, however, is only available to those who are mindful, who ponder, and who do not show pride. Water, the most precious commodity in the desert, gives life to the earth after its death. Muhammad promised, in effect, that there is a sign in the hydrologic cycle, which has water changing forms through evaporation, condensation, and precipitation, for those who listen.[57]

The Qur'an is not alone in providing geotheokolasis imagery involving water. In Haggai, the small, Old Testament book whose author claimed it was written in the second year of the reign of Darius the Great of Persia (ruled 522–486 BCE), God is heard ordering a drought on the land and on men.[58]

Whereas the water cycle is used by God as a means to bring forth life where there was none, he also uses the forces of violent storms to punish those who displease him. In Psalms 83:15, the author calls upon God to punish his enemies with fire, wind, and storm: "As the fire burneth a wood, and the flame setteth the mountain on fire. So persecute them with thy tempest, and make them afraid with thy storm."[59] A similar vision is found in the Qur'an:

When Allah intends evil to a people, there is no averting it, and besides Him they have no protector. He it is Who shows you the lightning causing fear and hope and (Who) brings up the heavy cloud. And the thunder declares His glory with His praise, and the angels too for awe of Him; and He sends the thunderbolts and smites with them whom He pleases, yet they dispute concerning Allah, and He is mighty in prowess.[60]

In a message that is reminiscent of the refrain heard in some Christian circles, "the Lord giveth and the Lord taketh away," the Qur'an tells of a man who did not spend his wealth in a way pleasing to Allah. Though the man's life was spent for the most part in lush environs, the fruits of his life's work were minimal and fell short of Allah's expectations. "Does one of you like that he should have a garden of palms and vines with streams flowing beneath it; he has in it all kinds of fruits; and old age has overtaken him and he has weak offspring, when, (lo!) a whirlwind with fire in it smites it so it becomes blasted; thus Allah makes the communication clear to you that you may reflect."[61] Muhammad described Allah as ready, willing, and able to take away that which he has given. The Qur'an's use of "whirlwind" in lieu of tornado is certainly in line with its usage in the book of Job 37:9, and the book of Isaiah 17:13 and 41:16. There is certainly a psychological imperative in these passages that does not permit a believer to become over-confident in his position relative to God. On the contrary, the fear of having storms of life descend upon one's house could well produce a persistent level of anxiety in the devoted follower of the prophets.

SACRED IMAGES OF WEATHER AND CLIMATE WITH INFERENCES FOR CLIMATE CHANGE

Moving away from images of geotheokolasis and geotheomisthosis, the sacred texts of the Abrahamic religions offer glimpses into past weather conditions and perhaps even changing climate conditions in the Near East. Weather imagery provides us with lenses through which we can see how the authors of those texts understood and coped with atmospheric conditions. As signs of God's pleasure or displeasure, weather images tell much about typical and atypical weather events that occurred over a span of time covering nearly two thousand years (ca. 1400 BCE to ca. 632 CE). Before discussing passages that suggest common or aberrant climatic conditions, it is helpful to start with some insight into average winter weather patterns for Jerusalem, Amman, Medina, and Cairo in recent years.

Most of the books that make up the Bible and Qur'an were written by people who presumably had a restricted knowledge of the world outside the Fertile Crescent. Aside from the Apostles Paul and Peter, few if any of the major authors ventured farther north than Syria. Paul and Peter moved about among eastern Mediterranean cities, but Muhammad most likely stayed in areas south of Syria, especially the Saudi Peninsula. It is hence probable that the writers of sacred texts were describing familiar atmospheric conditions. While Jerusalem has an average mid-January high of about 55°F and an average low of 36°F, one would not expect to find accounts of snowy or icy conditions, unless they were seen as aberrant. Snow does infrequently fall in the Near East, but any accumulation is short lived. Amman, Jordan, which is not far from the ancient land of Edom and the likely location for the home of Job, has the same average mid-January temperatures as Jerusalem. Medina, which is located on the Saudi Peninsula, is a good bit warmer in the depths of January; the average temperatures range from 54°F to 76°F. January is slightly cooler in Cairo, Egypt. Temperatures there range from an average low of 48°F to average highs around 64°F.

As one can see from the average January temperature ranges in those locations, it is likely that the authors of the sacred texts, if they experienced similar conditions, would regard icy, frozen conditions as unusual situations and not particularly applicable to their audiences who would not have had familiarity with them. In surveying the Bible and Qur'an for evidence of cold, snowy, or icy weather, the book of Job is rife with such imagery, and it even describes frozen bodies of deep water, which clearly indicates protracted cold spells. In Job 37:10, the author, who only a few believe was Moses (ca. 1500–1400 BCE), wrote, "The breath of God produces ice, and the broad waters become frozen."[62] In Job 38:30, a more general observation of cold weather is given: "When the waters become hard as stone, when the surface of the deep is frozen."[63] This same passage taken from the more poetic King James Version of the Bible states, "The waters are hid as with a stone, and the face of the deep is frozen." In Psalms 78:47, David described how God used hail to destroy Egyptian vines and frost to kill sycamore trees. David, who lived somewhere around 1010 to 970 BCE, was probably describing events that had taken place during the time of Moses, but his description of climatic conditions most likely reflected his own times (see chapter 11 for a discussion on climate change). This account of cold weather in Egypt was presumably passed on by Moses to subsequent generations. All references in the holy scriptures to frozen conditions are found in Job; the same is true for ice.[64]

Frost is mentioned in several Old Testament books: Exodus 16:14; Job 38:29; Psalms 147:16; Jeremiah 36:30; and Zachariah 14:6. In Exodus, Moses declares, "When the dew was gone, thin flakes like frost appeared on the desert floor." This passage is unclear as to whether the flakes were frost or they just looked like frost. Nevertheless, Moses's use of the word suggests that his audience would have been able to recognize frost and were accustomed to seeing it. In Job, the author rhetorically asks, "From whose womb comes the ice? Who gives birth to frost from the heavens?" In Job 38:22–23, the author asks, "Hast thou entered into the treasures of the snow? Or hast thou seen the treasures of the hail, Which I have reserved against the time of trouble, against the day of battle and war?"

In the book of Jeremiah, it seems the author understood that snowfall in Lebanon was sufficient to recharge aquifers. The quality of the water must have been exceptional because Jeremiah asked, "Will a man leave the snow water of Lebanon, which comes from the rock of the field [likely bubbling out in springs]? Will the cold flowing waters be forsaken for strange waters?"[65]

About one hundred years before Jeremiah penned his book around 600 BCE, Isaiah wrote about snow in such a way that suggests it was usual and part of the water cycle that provided life to plants and farmers: "For as the rain comes down, and the snow from heaven, and do not return there, but water the earth, and make it bring forth and bud, that it may give seed to the sower and bread to the eater."[66] In the time of Solomon, around 900 BCE, the author of Proverbs was so clearly convinced that snow was typical of winter months that he used it in a metaphor to show the incompatibility of honor and foolishness: "As snow in the summer and rain in harvest, so honor is not fitting for a fool."[67]

The pattern of weather-related examples in the holy texts, which were penned over an expansive period of time, suggests that climate change has indeed occurred in the Near East. Applying evidence and concepts from archeology and meteorology to the discussion strengthens that contention. For example, Jesus' ministry was focused along the shores of the Sea of Galilee. At least three of his disciples (Phillip, Andrew, and Peter) lived in a Galilean fishing village named Bethsaida ("the house of the fisherman"). According to Israeli archeologist Rami Arav and an international team of anthropologists, they believe that they have found biblical Bethsaida. However, it is located two kilometers from the northeastern shore of the lake.[68] Given the lake's shrinking size, it would seem that the regional climate is becoming drier and perhaps hotter than it was in the first century. In addition to drier winters, the Sea of Galilee suffers from human consumption of its waters. At over two hundred meters below sea level, the fresh water lake is endangered of becoming irreversibly saline, which was the fate of the Dead Sea.

As we will discuss in chapter 9, tornadoes or whirlwinds, as they are called in the Abrahamic texts, occur in a only a few places on the planet; yet Job 37:9 describes weather conditions that produce hail and spawn tornadoes or whirlwinds: "Out of the south cometh the whirlwind: and cold out of the north." The subsequent verse declares, "The breath of God produces ice, and the broad waters become frozen." These verses suggest that colliding masses of southerly, moisture-laden air and northerly, cold, dry air were common occurrences when that holy book was written. Further evidence of the regular occurrence of intense frontal activities in the time of Job can be found in 38:22: "Hast thou entered into the treasures of the snow? Or hast thou seen the treasures of the hail." In more recent versions of the Bible, "treasures" appears as "storehouses," so it should be construed that climate during the time of Job was colder and wetter. Why else would Job's neighbors need storehouses for times of snow or hail? Clearly stormy and snowy conditions were so common that it was necessary to put away food and provisions for such times. As is known today, it is in these frontal situations that hail is produced and tornadoes are formed. However, hail and tornadic activity are now oddities in the eastern Mediterranean. In 2006, a mini-tornado was spotted in western Galilee. It produced damaging hail and uprooted a few trees. Baruch Zvi, a climatologist working at Tel Aviv University, stated that "an event of this kind is very rare in our region. The last time a tornado took place in the area was in neighboring Cyprus. A few mini-tornadoes," he recalled, "took place on the coast and in the Sharon area in the past."[69] As was shown earlier, Muhammad also described whirlwinds, which would most likely have been caused by contrastive air masses. Given the effective description of the weather conditions that give birth to snow, hail, and tornadoes in the book of Job and their infrequency in the region today, it is logical to assume that climate change in the Near East has taken place. The larger and elusive question is "Why?" That discussion, however, will be picked up in chapter 10.

SUMMARY

The geotheological imagery presented in this chapter is by no means an exhaustive delineation of the ways in which prophets in the Abrahamic religions viewed the interplay between nature and the worship of God. On the other hand, the writings of the prophets collectively provide an overview of the basic ways in which ancient peoples in the Near East understood human-environmental interactions. The descriptions of weather conditions certainly enable us to see that extended cold, wetter weather occurred when the oldest texts were written. Furthermore, it is important to note that they are absent in the more recent writings.

The concept of geothology is a significant aspect of the ancient Near East. As the point of origin for all non-African humans living on six of the world's seven continents, elements of geotheology have made their way into the secular debate over climate change, which is a topic we will discuss in chapter 8. While the lexicon of John K. Wright is usefully applied to the study of human-environmental interactions in ancient times, it does not account for the ways in which God is seen using nature to punish and reward human actions. Geotheokolasis and geotheomisthosis can help fill that void in Wright's taxonomy. Moreover, since many secular people still regard nature as capable of punishing and rewarding humans for their abuse or wise management of natural resources, geokolasis and geomisthosis should likewise be added to the geographic lexicon. While it may seem that these concepts suggest a conscious awareness on the part of nature to administer punishments and rewards, such is not the case. While natural laws do not direct human behavior, the laws of nature dictate natural forces. Benefits and adverse consequences resulting from human interactions with natural forces are produced by living within or outside the narrow parameters of the laws of nature. As is shown in later chapters, there are laws in nature that, if we do not take them into account as we go about our daily lives, can result in adverse and even deadly consequences.

As the passages illuminated in the later sections of this chapter demonstrated, snow and icy conditions were observed by the author of the book of Job (ca. 950 BCE); snowy conditions are described in Isaiah (ca. 700 BCE) and Jeremiah (ca. 600 BCE). From the book of Daniel (ca. 550 BCE) onward, most references to snow are used as an example of cleanliness or purity as in "white as snow."[70] Interestingly enough, frost disappears from the scriptures after Jeremiah. There are no accounts of snow (as a precipitation event), icy conditions, or even frost in the New Testament or in the Qur'an. Judging from the sacred passages, it seems that the climate was much cooler and wetter in the time of Job, and then a gradual warming trend followed, perhaps intermittently, so that by the time of Darius (around 550 BCE), snow, hail, and frost imagery are lost from the texts. This certainly does not mean that frost or snow never fell again in the Near East, it simply suggests that their impacts must have been light because little or no attention was given to them. It is interesting that lightning, thunder, whirlwinds and other meteorological events are seen as divine punishments, but snow and ice are depicted simply as the result of God's will. They are not generally seen as forms of punishment. Indeed, they are depicted as sources of life-giving water, bubbling up from the fields during the warmer and drier months of the year. When snow is mentioned in the later books of the Bible it is often in reference to purity. Biblical images of ancient times along the southeastern shores of the Mediterranean Sea reveal that climate change has occurred even during the Holocene. The possible return of a mini ice age, or even a major ice age, should force us to rethink climate change research to include "what if?" scenarios for a much colder climate. If rivers in Israel were routinely frozen, one can only imagine the drop in temperatures and the accumulation of snow and ice in places like Buffalo, New York, and Berlin, Germany. I know of no human-made structures that can support one-mile-thick layers of ice on their rooftops. Indeed, if glaciers can scrape out sections of earth to form the Finger Lakes in New York or build linear hills called moraines in southern Michigan, they would pulverize anything made by human hands.

Chapter 5

GEOKOLASIS

The Cases of Donora, Lake Erie, and the Dust Bowl

INTRODUCTION

A fear of geokolasis, or the deleterious impact of nature in response to human actions, is sufficient enough to ignite a social movement that calls for a return to more rustic lifestyles close to nature or, as some companies that hope to capitalize on the concern call it, "go green." The United States is certainly not alone in the green movement. In the United Kingdom, a place where seventeenth-century intellectuals presumed to make mutually agreed upon covenants between their nation and God, has seemingly made a similar pact with nature. Whereas seventeenth-century Puritans believed that God would send famine or some other dreaded condition as a means of reckoning, the government, apparently out of a fear of nature, has tied punishments and rewards to its "People and Planet's Go Green Campaign."[1] While many of the initiatives stemming from the green movement are well founded, it is likely that any dreaded geokolasis and the illusionary promise of geomisthosis produce myths to support these organizations' premises, including some that are not based on solid science. One particular concern of mine is a repeat of the rural renaissance of the 1970s in which the pursuit of being close to nature led to the conversion of millions of acres of crop and forest lands into housing developments and later asphalt-covered parking lots for schools, hospitals, and other service-sector businesses.[2] In other words, going green must be pursued with caution because presumed solutions may actually create new and unexpected environmental consequences.

It is important to abate the fear of geokolasis and replace it with a desire to apply logical analysis to human-environmental interactions. This chapter will show that shifts in global investment patterns may well be responsible for ameliorating twentieth-century environmental crises in the United States—not necessarily environmental policies and their implementation. This view is supported by the fact that industrialization has risen in developing countries in Asia and Latin America while manufacturing in the United States has declined. Meanwhile, as will be shown in this chapter, air and water pollution conditions in the United States have improved. However, many rapidly industrializing countries in Asia are repeating many of the West's environmental mistakes. Developed countries like the United States demand cheap products manufactured in countries that do not have adequate environmental safeguards, and are hence culpable in continued pollution scenarios. However, it is a mistake to believe that because the United States and other Western countries buy Asian products they can profoundly influence the adoption of environmental policies in Asia. Certainly, rising levels of economic development among Asian countries likewise minimizes even the United Nations' ability to influence Asian nations to adopt regulations that would prevent repeats of environmental calamities that reduced the quality of life in Europe, Japan, and North America during their respective Industrial Revolutions.[3] Because Asian manufacturing and wholesale companies owned by Chinese, Malaysian, Japanese, Korean, Bangladeshi, Indonesian, and other corporations based in ASEAN (the Association of Southeast Asian Nations) countries are now selling to their Asian neighbors, the weight of Western sentiment as it might manifest itself through product boycotts aimed at influencing environmental policies in the East is greatly reduced.

It is arguably hypocritical for Western countries to tell developing nations that it is important to control their economic growth and, hence, resource exploitation to safeguard themselves and indeed the world from potential environmental calamities. This thought is strengthened by the realization that populations in southern and East Asia are growing much faster than their counterparts in the developed world. Countries located in southern and eastern Asia have incredible demographic pressure on them to expand their carrying capacities, and economic growth and diversification, aside from coercive, draconian policies, are the only proven ways to reduce population growth rates. This assessment necessarily assumes that economic diversification encourages urbanization, widens social options for women, and lessens the value of having large families. These patterns are clearly seen in the discussion on demographic transitions in chapters 11 and 12. Countries including Italy, Russia, Germany, and Japan epitomize the impact that economic development and women's empowerment have on population growth. Their crude death rates now exceed their crude birth rates. Without immigration to replenish their human resources, they will eventually die out, albeit centuries into the future. At some point, the political power of immigrant populations in those countries will exceed that of charter residents. When that happens, Germany and Italy, which heavily rely on immigrant labor, may well go the way of the Roman Empire when its power structure was taken over by non-Romans from central and northern Europe. In a political context, Germany and Italy will likely collapse from within.

ASIA FILLS THE PRODUCTION VOID

Global shifts in investments in factories and the tools of production accelerated in the 1970s partly because of increasing costs of labor and other production inputs including new technologies to reduce emissions and effluents. Asian countries seized the opportunity to fill the production void. In 1967 and in response to mounting population pressures, ASEAN was formed by Indonesia, Malaysia, the Philippines, Singapore, and Thailand. Membership has since expanded to include Brunei, Myanmar, Cambodia, Laos, and Vietnam. The organization is dedicated to accelerating economic growth and social progress among its member nations. ASEAN countries are not alone in responding to their own challenges to expand their respective carrying capacities via innovation and economic growth. Indeed, China, which still generates 80 percent of its power from coal, seems bent on charting its own course despite clear evidence of atmospheric coal-based emissions from its industrial plants.[4] Five of the world's top-ten most polluted cities are in China. They are Beijing, Shanghai, Shenyang, Xi'an, and the ancient city of Guangzhou.[5] While pollution is certainly a problem with which to contend in those heavily settled areas, the relative rise of a new urban middle class in Asia has increased markets for agricultural imports, which has a positive impact on America's trade imbalance.[6] However, as is shown shortly, connecting markets to points of production may be economically expedient in the short-term, it gives rise to higher levels of gaseous pollutants while simultaneously shifting the balance of power in the world. Where money is concentrated, political power will be as well.

As countries in Asia fill the production void created by the decline of manufacturing industries in the West while selling to global markets, atmospheric pollution levels, despite abatement technologies, are increasing. One of the by-products of production is sulfur dioxide (SO_2), which is a pollutant frequently analyzed by atmospheric scientists because of its "strong regional effects, available abatement technologies, and different regulations across countries. Moreover, a deeper understanding of SO_2 emissions contributes to a better understanding of three [related] environmental problems: air pollution and smog, acid rain, and global climate change."[7] Given that about 45 percent of SO_2 is emitted through the production process, it is logical to assume that changes in manufacturing around the world would be reflected in regional and global levels of SO_2. As Western factories tooled down and Asian factories tooled up, there was a drop of about 3 percent in global SO_2 levels between 1900 and 2000. Abatement technologies certainly helped reduce an unknown quantity of SO_2 levels, but as marketing efforts improved, transporting goods between production points in Asia and consumer markets around the globe are raising SO_2 emissions.[8] Shipping food items grown in the United States to hungry urban markets in Asia also contributes to rising SO_2 levels.

Beyond the issue of pollutants' levels in the atmosphere, the evolving economic system suggests that humans are adapting to the forces of nature by expanding the earth's carrying capacity. By shifting points of production to regions with high population densities, impoverished masses are less likely to take part in huge population flows that have characterized human history. The evolution of human adaptation seems to have followed this trajectory: our earliest ancestors left familiar environs in search of flora and fauna; succeeding generations moved from rural to urban places in search of jobs; while people are still migrating in large numbers, the shift in points of production from places with static or declining natural increase rates to areas of high population growth is, at least for the present, a more efficient and expedient way to fit more people into the ancient ecumene.

Nevertheless, some observers are hopeful that the Chinese and ASEAN governments will adapt as the West has to environmental pollution. However, it must be admitted that while environmental protection policies and resulting abatement technologies have been enacted and implemented in the West, divestments in industrial production in the American Rust Belt and Europe are more likely responsible for improvements in each region's air, land, and water qualities. It is as if pollutants emanating from manufacturing processes are like the contents of a tube of toothpaste. If pressure is applied on one end of the tube, then the paste shifts to the other end. When Western governments applied environmental pressure on domestic industries to lower their levels of pollutants, manufacturing activities declined in the West, but within a short time, they reappeared in Asia and Latin America, where governments eager to feed their growing populations allowed manufacturers to operate inside their borders. It is now up to Asian and Latin American peoples to respond to their own emerging environmental crises. How will they adapt? Will declining standards of living in the West lead to a return of higher rates of fertility in heretofore developed countries? Will there be growth in the number of poor people living in the West, and will their governments reduce environmental protection standards to attract manufacturing jobs? Or, conversely, will global patterns of wealth and demographic transition stabilize? With that positive thought in mind, let us explore ways in which the United States created and then responded to environmental problems that have manifested themselves in urban air pollution crises, the "death and rebirth of Lake Erie," and the Dust Bowl. As these discussions show, geokolasis has occurred—and undesirable situations like these can return and plague us once again.

DEATHS IN THE FOG

On October 26, 1948, a Pennsylvania high school football rivalry was about to be rekindled. Cheerleaders and band members from Donora and Monongahela High Schools performed before a packed stadium as they waited for their teams to take their respective places on the field. Nestled in a horseshoe bend along the Monongahela River, the town of Donora sat surrounded by hills rising some 350 feet above the playing field. Frequently occurring high-pressure centers and the local landscape features encouraged the retention of pungent smells of emissions belched out from steel mills and zinc-smelting plants that served as sources of income for many families in the area. During the pregame activities, the evening seemed to be no different from others in the past. The smells that enveloped the football field were familiar to players and others in attendance, but because of an unfortunate interaction between human activities and natural forces, this night *was* different.

Unbeknownst to players, coaches, and team supporters, a temperature inversion was forming over the settlement of twelve thousand mostly working-class people. After sunset, cool air descended the hills around the steel town. Warmer air was pushed aloft and formed a ceiling, trapping emissions near the ground. Relatively cool particles emitted from manufacturing plants served as condensation nuclei on which water droplets formed. A multicolored fog settled in over the football field and the surrounding area. By halftime, the game had to be cancelled because the players complained of labored breathing. High levels of sulfur dioxide, nitrogen dioxide, and hydrocarbons were measured in the atmosphere, while soot from local smokestacks dusted everything exposed to the foul night air. Like a freshly fallen snow, sidewalks had footprints on them. Many of them were left by people scampering home to escape the ominous fog that had swallowed Donora.

The fog stayed over the town for six days. During that time 5,910 residents became ill with symptoms that included nausea; vomiting; severe headache; nose, eye, and throat irritation; as well as labored breathing and respiratory constrictions.[9] At four o'clock on Thursday morning, Iven Ceh, a seventy-year-old immigrant from Yugoslavia, became the first to die from respiratory complications. But by the end of "Black Saturday," the number of deaths had reached seventeen. Two more people died on Sunday, and another died later, bringing the death toll to twenty. As of 1948, this was the worst air pollution disaster in the United States, but its distinction did not last long. The death toll in Donora paled in comparison to that in New York City during episodes in 1953 and again in 1956.[10]

On November 12, 1953, a stagnant high-pressure center descended over the nation's largest city. It lingered over the Big Apple for ten days. Sky-darkening soot, dust, and fly ash accumulated in the air above the city to an average concentration of 3.5 tons per cubic mile.[11] The city had a measured sulfur dioxide level that was five times its normal level. Airports were shut down due to poor visibility. Ferry trips to the Statue of Liberty were also cancelled due to the pervasive and unrelenting smog. Tens of thousands of people experienced burning eyes and uncontrollable bouts of coughing, but those were minor consequences of the unwelcome shroud that cloaked the city. By the time atmospheric conditions improved on November 22, some two hundred people were dead.[12] As if that weren't enough, merely three years later, an even more devastating event occurred in New York City. In just an eighteen-hour period, at least one thousand people died from smog-related illnesses.

The industrialized United States was not alone in suffering from smog-related deaths. London, a city famous for its "fogs," also suffered from the wrath of nature. On December 4, 1952, an extradense "pea souper" settled in over a seven-hundred-square-mile area of greater London and stayed in place for five days. Visibility was so poor that men carrying lit flares led crawling buses through the city's streets.[13] Theater performances were cancelled due to poor visibility. Thousands of Londoners flooded hospitals with symptoms ranging from labored breathing to cyanosis. By the time the temperature inversion lifted and drier air from the north further dispersed the pollutants, at least four thousand people had died from heart disease, bronchitis, and bronchopneumonia—all of which were attributed to the smog.[14]

Those events and the empirical data associated with them led to technological innovations—primarily smokestack scrubbers and increased combustion temperatures—to ensure oxidization of particulates, carbons, nitrogen dioxide, and sulfur from emissions. Nitrogen dioxide and sulfur are acid precursors, which means that when they are combined with oxygen and water in the atmosphere, they form sulfuric and nitric acids. These acids fall to the earth in rain droplets and snowflakes where they acidify lakes, kill or harm vegetation, and accelerate deterioration of buildings and statues. The spillover costs associated with these deleterious impacts have forced the governments of Canada, the United Kingdom, and the United States to pass a number of acts to regulate and control emissions. Great Britain has even shifted heavily to nuclear power generation. City governments have also regulated the burning of household wastes because in Rust Belt cities like Detroit, even as late as 1968, residents burned household refuse, including plastics, which release carcinogens when burned, in allies behind their homes.[15] Nevertheless, these changes in behavior or adaptations were the result of societies interfacing proactively with environmental limits.

Today, Donora and the greater Pittsburgh area have remarkably clean air. However, heavy manufacturing has also declined in the region, so it is difficult to assign credit in improved air quality in this part of the Rust Belt to technology or to shifts in global investments in coal-fired steel production. Because environmentally safe technology is expensive and would make their products less competitive in a global market, Asian countries have been slow to enact laws to safeguard their respective environments. Nevertheless, the technology to improve the quality of air in highly industrial places is solid and reliable, so there is hope that China and ASEAN nations will adopt more efficient ways to produce energy to run production facilities.

We began this chapter with air pollution in urban places, but now let us focus our attention on lake habitats and water consumption in rural areas because environmental problems are not exclusively found in densely populated settlements.

"DEATH AND REBIRTH OF LAKE ERIE"

Despite human abuses, nature is resilient. In the face of tons of phosphates, nitrates, and a host of solid waste pollutants poured into it, Lake Erie is an amazing body of water with a vibrant ecosystem that offers billions of dollars of economic utility. Fishermen today enjoy reeling in black crappie, Chinook salmon, rainbow trout, walleye, and an assortment of bass. While this may seem an insignificant pastime for many, it is something that a number of environmentalists, as late as in 1980, did not think was possible on Lake Erie. Indeed, many popular writers in the 1970s and early 1980s regarded Lake Erie as dead.[16] While human actions caused changes in the composition of the life-forms that depend on the lake's ecosystem, it was not dead. In fact, the amount of biomass supported by the lake in the 1970s was the same as it was in the 1920s, roughly fifty million tons.[17] Upon closer inspection, it was the economic value of Lake Erie that had died and then was reborn. Its future, however, is not assured; some of the less obvious source inputs that caused its economic demise are still present and represent challenges to those agencies charged with protecting the lake's complex ecosystem. However, the decline in the amount of contaminates entering the lake from the Detroit area via the Detroit River is arguably more a function of a reduction in heavy manufacturing in the region than it is from the implementation of policies, so it is doubtful that over the next few decades, the lake will be forced to absorb pollutants at pre-1980 levels.

In the decades prior to the 1920s, Lake Erie supported a multimillion-dollar commercial and sport fishing industry. The lake's commercially valuable species included cold-water-demanding lake trout, blue pike, cisco, whitefish, walleye, and sauger. These fish not only need cold water, but they also require high levels of dissolved oxygen. In the warmer months of late summer, these fish species congregate in the cooler, oxygen-rich deep waters of the hypolimnion (bottom layer of water). Fifty years of farming and industrial development set the stage for a costly environmental calamity. As air pressure gradients have stabilized along with increased atmospheric temperatures since 1980, the mixing of water layers has declined. Normally, atmospheric instability produces wind that, in turn, causes water molecules to move in response to atmospheric friction. This movement ensures that oxygen will be mixed throughout vertical profile of the lake. Ironically, stable weather actually hurts the lake's ecosystem. It also makes it difficult to determine the direct impact that humans have on fish. Nonetheless, dissolved oxygen levels do impact the biological processes that threaten the productivity of the lake.

In addition to changes in climate and weather, Lake Erie's ecosystem was further altered by human activities that spurred on eutrophication, which is a process of enriching a stream, pond, river, lake, or sea with nitrates and phosphates. Eutrophication encourages high growth rates among plant species, including weeds and short-lived algae. Of primary culpability in Lake Erie's ecological crisis was waste material discharged into the lake. For decades, Cleveland, Detroit, Toledo, Erie, and other cities that sit on or near the shores of Lake Erie allowed untreated and partially treated wastes to runoff into the lake, although 80 percent of the water flowing into the lake washes the shores of Detroit and Windsor, Ontario, before emptying into the lake as part of the Detroit River.[18] During the peak years of pollution, there were nearly eleven million people contributing refuse to the collection of materials enriching Lake Erie.[19] Furthermore, in the 1970s there were still some thirty thousand acres of agricultural lands bordering the shoreline, so annual rainfall and snowmelt washed tons of nutrient-rich manure, urine, and commercial fertilizers into Erie. There were also chemicals such as herbicides and pesticides that were streamed into the basin. Third, nitric oxides expelled from millions of cars that operated in the lake's watershed were carried over it, where water molecules and oxygen interacted with the chemical to form dilute nitric acid. Nitric acid then also washed into the lake. There it formed nitrates, which further enriched the water and spurred on eutrophication. The last major cause of eutrophication was nature itself. Minerals and nutrients from rocks and soils in the watershed have always enriched the lake. Unfortunately, land development and farming sped up the erosion and deposition of naturally occurring nutrients.[20]

The process of eutrophication that plagued Lake Erie produced large algae blooms; particularly troublesome was and continues to be the filamentous algae *Cladophora*. Being short-lived, dead and dying *Cladophora* had, by the 1970s, accumulated in the hypolimnion to a thickness of 30 to 120 feet. The nutrient-rich ooze on the lake bottom lay dormant for decades. Its dormancy was facilitated by relatively high levels of dissolved oxygen in the hypolimnion, which encouraged the formation of a sealant on the ooze. Oxygen levels were sufficient to form ferric hydroxide, an iron-rich compound that preserved the ooze, trapping its nutrients on the bottom of the lake. However, when the level of oxygen in the hypolimnion declined, the insoluble ferric hydroxide was converted to soluble ferrous phosphate, and the nutrients were then released to literally fertilize further algae growth. The decomposition of dead algae was facilitated by oxygen-demanding bacteria that lowered the amount of dissolved oxygen.

In late summer 1953, natural conditions set that economically calamitous scenario in motion. A protracted high-pressure center settled in over the lake and stayed in place for twenty-eight days. In the shallow western basin, temperature regimes became rigidly stratified. With no winds to mix layers of water, oxygen levels in the hypolimnion dropped precipitously. As a result, the insoluble ferric hydroxide was converted into ferrous phosphate and the fuel required for enhancing eutrophication was unleashed. Mayflies, which are an important prey species for commercially valuable fish like white fish and lake trout, were forced to leave or die from asphyxiation. Valuable fish either left or were otherwise killed by a lack of oxygen or by decreases in their food supply.[21] This situation had a tremendous impact on fish harvests. Whereas in 1920 some fifty million pounds of fish were caught, in 1970 only one thousand pounds of fish were harvested. This represents a decrease of 99.99 percent.[22] In response to concerns about pollution as well as declines in fish harvests, the United States and Canada entered an agreement in 1972 called the Great Lakes Water Quality Agreement (GLWQA). The GLWQA was renewed in 1978 and amended in 1987.[23]

Up to the mid-1990s, improvements were measured in the amount of dissolved oxygen in the central basin's hypolimnion, and sediments from tributaries had been reduced by 50 percent.[24] While there has been some reversal of those trends in recent years, the lake is once again the home of commercially valuable game fish such as black crappie, Chinook salmon, rainbow trout, walleye, as well as white, small, and large mouth bass. Jet Skis® and swimmers are also abundant. There is, however, concern that the improvements gained over the last three decades could be lost through declining public concern and enforcement of emissions standards. The fact that declines in auto production and output from other industrial plants located along the Detroit River have occurred simultaneously with improvements in the lake's health may well be part of the reason that Lake Erie is now tenuously reborn. To be sure, even popular writers who tend to fear for the future of Lake Erie note that there are measureable declines in heavy metals and PCB (polychlorinated biphenyl) contaminants.[25] As a result of the uncertainties with assigning credit for the lake's improved health, the jury is still out on the effectiveness of policies designed to reign in life-threatening pollutants that, although lessened, still plague Lake Erie's ecosystem.

There certainly are writers who are concerned about geokolasis if serious steps are not taken to further protect Lake Erie and, indeed, all of the Great Lakes from harmful abuses inflicted on them by people. According to Ohio journalist Chris Evans, "Beneath the wake of Jet Skis and barges, an apocalyptic environmental stew simmers in the insidious heat created by greenhouse gases. . . . It is a harbinger of doom. When Lake Eire, that Lazarus of the Great Lakes, was declared dead back in the 1960s, the first visible sign of sickness were blue-green algae blooms birthed by high phosphorous levels. Now 'they're baaaaack.' And some of them are toxic."[26] Evans is optimistic, however, because President Obama and the US Congress are working toward passage of the Great Lakes Restoration Initiative, which will provide up to $475 million to target invasive species, ensure habitat conservation, and reduce pollution. This would be the largest cash infusion from the federal government into the fight to protect the Great Lakes.[27]

Ironically, Lake Erie's improved health and the host of environmental problems associated with industrial emissions in Donora and New York City, unlike the causes and cures for the Dust Bowl of the 1930s, may actually be the result of shifts in global investment patterns and resource exploitation and not necessarily through direct policy implementations.

GRAPES OF DRAFT

As most high school students in the United States know, author John Steinbeck won a Nobel Prize for Literature in 1962 for *The Grapes of Wrath* (which itself won a Pulitzer Prize in 1940), a novel about displaced Oklahomans resettling in California's Central Valley. While Steinbeck was determined to tell the human story of the consequences of reaping the wrath of what had been sown in the Great Plains, especially in western Oklahoma, he gave us a sobering glimpse of the primary causes of the Dust Bowl that had set the migration of Oklahomans in motion. It is helpful to recall from chapters 1 and 2 that humans have always migrated in search of better environs and less competition for resources. This situation was no different for Steinbeck's Okies, and it was certainly no different for the Tennesseans and Kentuckians who had become Oklahomans a generation or two earlier. Let us not forget that the Tennesseans and Kentuckians had been Pennsylvanians and Virginians before settling west of the Appalachians, and before they were Pennsylvanians and Virginians, they were Irish, Scottish, German, English, Asian, and African folk. With new rural places to settle and carry on with an agrarian life as they knew it, there was no reason to change or adapt. Such was the case with the Okies and their migrating ancestors.

When northwestern Europeans first arrived in North America in the seventeenth and eighteenth centuries, they brought limited farming knowl-

Image 5.1. California's Central Valley. *Image courtesy of the author.*

edge with them. The Germans were perhaps the best farmers, having already fabricated effective plows for busting clods of earth. The Scots and the Irish were primarily pastoral peoples, and for safety reasons, they often chose to settle in upland places that did not offer deep soil for farming. Africans and English folk had learned to deep plow in the coastal plains where cotton and tobacco were grown. The Cherokee, Creek, Chickasaw, Choctaw, Seminole, and Shawnees who preceded them to Oklahoma used crude farming implements. North of Oklahoma and before white settlement on the Great Plains, Mandans, Arikaras, and Hidatsas lived a sedentary lifestyle along the Missouri River. They raised produce grown on modest farms in the river's fertile bottom lands.[28] Collectively, the people who settled in Oklahoma and the Great Plains were not well-informed farmers. There were no traditions of contour farming, crop rotation, or no-till agriculture, and they certainly had no idea that chelating agents (bonding agents to hold soil nutrients in place) were present in the root systems of legumes such as clover, alfalfa, peanuts, or soybeans. Ranchers in the post–open range days also had not determined that grazing lands could support only a limited number of cattle (i.e., one cow per acre). Like their ancestors, when game or land fertility played out, ranchers picked up their belongings and moved westward to better lands. By the time Oklahoma became a state in 1907, the continent had been settled and there were few places in which an independent-minded farmer could resettle without someone already laying claim to that coveted space—a situation that Oklahomans found in California's Central Valley. In fact, the term "Okie" was originally a derogative label placed on poor people driven off the Great Plains by the wrath of nature. With nowhere else to go, North Americans had to adapt to the settlement circumstances that presented themselves. The Dust Bowl of the 1930s proved that point.

As already alluded to in the opening paragraphs of this section, the seeds of the Dust Bowl were planted thousands of years earlier, in the human practice of migrating away from places plagued by too many people competing for too few resources. The Great Plains were the last American region to be settled. California had become a state in 1850. Oregon joined the Union in 1859 and Washington in 1889. The United States government actually had to offer incentives to get people to settle in the Great Plains because in the decades prior to the passage of the Kansas-Nebraska Act (1854) and the Homestead Act (1862), which featured allotments of 160 acres per claim, the treeless plains had little appeal. It was assumed by many that the absence of trees was a clear sign that prairie soil was unfit for farming. Trees were also a resource needed for fuel and for building homes and barns. It is no small wonder why the Oregon Trail cut *through* the Great Plains on its way to the forested lands of the Northwest. These negative attitudes toward the Great Plains were certainly reinforced by the notion that the interior of the continent was known as the "Great American Desert." That rather dismal and wholly inaccurate name was assigned to the region in an 1823 report by Major Stephen H. Long. However, the seeds for the "unsuitable for cultivation" assumption about the plains were planted nearly two decades earlier. From reports made by such figures as Meriwether Lewis and William Clark, who explored the area between 1804 and 1806, and Zebulon Pike (1806–1807), easterners developed the idea that the continent's inner parts were incapable of commercial farming.[29]

The Homestead (1862) and Kansas-Nebraska (1854) Acts attracted thousands of families onto the Great Plains. Within weeks of the passage of the Homestead Act, Oklahoma Territory climbed from a few hundred non–Native Americans to approximately sixty thousand settlers. It was a seemingly mad rush to claim parcels of 160 acres per allotment. President Lincoln, it could be argued, was punishing the Cherokee and other regional tribes for having supported the Confederacy by taking their lands and issuing them to non-Natives.[30] Indeed, the Army of the Trans-Mississippi, a major Confederate force, was made up of mostly of Cherokee, Muskogee, and Seminole tribesmen. It was formed after Chief John Ross and the Cherokee Council decided to join the Confederacy. Under the command of General Stand Watie, notables such as Cherokee senator Clement Vann Rogers, father of the humorist Will Rogers and my cousin, served as an officer in the lost cause.[31]

Encouraged by a vengeful federal government, white settlers soon outnumbered the eastern tribes that had arrived there in the late 1830s. From 1860 to 1910, 234 million acres passed from public to private ownership. By

Image 5.2. California's giant Redwoods. *Image courtesy of the author.*

Image 5.3. The author by a Redwood tree. *Image courtesy of the author.*

1900, the population of settlers who supported themselves as farmers or ranchers was nearly forty thousand.[32] In 1860, Oklahoma was known as Indian territory, but in 2000, only 7.9 percent of the state's population was Native American, and another 4.5 percent claimed a mixed race identity.[33] Lincoln's initial policy toward the Native tribes that called Oklahoma their home not only destroyed sovereign nations, but it also introduced settlers armed with farming technology, including the knowledge to use it, that was not suited to a part of the country that had limited agricultural potential. The relatively drier climate in the central and western portions of the state was the primary limiting factor in developing a commercial agricultural economy.

Kansas and Oklahoma's landscape and vegetative cover change remarkably from east to west. The elevation also steadily climbs in the western reaches of the Great Plains. In describing the relative flatness and treelessness of western Kansas, a local man jokingly told me that, not long ago, he had "watched his dog run away for two days."[34] Western Kansas and Oklahoma are indeed vast, open spaces. For people accustomed to urban environments or hilly, forested areas, the sensation of dry prairie winds blowing across the skin leaves a lasting impression.

The natural features of the southwestern Great Plains, arguably the hardest hit during the Dust Bowl of the 1930s, include fertile, humus-rich chernozem and mollisol soils, but, as a region, it routinely experiences periods of drought. Stretching from Texas, through eastern Colorado, and into western Oklahoma and Kansas, precipitation amounts total less than five inches per year during severe droughts. This is a major problem for farmers and ranchers because about twenty inches of precipitation are annually needed to raise marginal crops. From what is known from the geological record as well as from oral and recorded history, these droughts seem to occur in twenty-two-year intervals.[35] Although river bottoms offer ample ground-based moisture for small stands of cottonwood, hackberry, dogwood, scrub oak and conifers such as eastern red cedar and common junipers, as well as other deciduous trees like maples and tulip trees, the dry climate of the open prairie makes it difficult for trees to mature.

Unquestionably, drought is an unwelcome visitor in any agricultural area, but, in the southwestern plains in the late 1800s and the first quarter of the twentieth century, dry spells came more frequently and stayed longer. In 1890 and again in 1910, drought struck the region with devastating results. Crops withered away, and a number of farms and ranches were temporarily abandoned. These meteorological events were insignificant in comparison to what happened as a result of the drought that lasted from 1926 to 1932. Annual rainfall was barely sufficient to settle even minor dust devils, let alone raging dust clouds. Unbeknownst to farmers and ranchers, they had inadvertently prepared the soil for the formation of dust clouds that produced muddy rain as far east as Washington, DC. Oklahoma prairie soil settled on the decks of a cargo ship two hundred miles out in the Atlantic.[36] Dust particles from the Great Plains made their way into clean luxury apartments and offices in New York City.

Meanwhile, back on the Plains, blackened clouds or black blizzards formed during wind storms, blocking visibility and creating hazardous health conditions. In March 1935, microscopic particles kicked up in a Colorado storm caused forty-eight relief workers to contract "dust pneumonia." Four of the workers died from the illness, and in another dust storm in New Mexico, five children in one family were smothered to death in their beds.[37] On May 11, 1934, one storm displaced three hundred million tons of fertile soil, which, interestingly enough, is the equivalent amount of the earth scooped away to dig the Panama Canal.[38] These black blizzards were not caused by nature alone.

In the open prairie of the southwestern plains, grama, big blue stem, and little blue stem, as well as buffalo grass, provide natural protection against erosion. Their intricate root systems house nitrogen-fixing bacteria along with chelating agents that bind soil material. Natural conditions had encouraged healthy soil structure and herbaceous plant productivity before the territory was settled by farmers and ranchers in the 1850s and 1860s. While plows broke up root systems on farms, heavy cattle grazing packed down rangelands, effectively reducing soil structure. Overgrazing, too, was a contributing factor in decimating grass clumps. With no root systems in place to keep soil from blowing away, the stage was set for the Dust Bowl. As Will Rogers described the activities that led to the Dust Bowl:

Any old great civilization that they've ever gone and dug up, they go down and find two or three layers all buried there in the dust. And they've all been covered up because of plowed up ground that shouldn't ever been plowed up in the first place and the trees cut down that shouldn't never been cut down anyway.... Now, remember a couple of years back when [President Franklin D.] Roosevelt suggested plantin' millions of trees all across the dry regions? He said, "Ever so many miles, we'll put a row of trees clear across the country!" Well, the Republicans had one of the best laughs they've had since 1928 when they read that. "Imagine the government going into the tree planting business! What a nut idea!" Well, it was so nutty that it would be about ten to fifteen years before they'd be compelled to do it—that's how "nutty" it was![39]

Along with the aviator Wiley Post, a fellow Oklahoman, Will Rogers was killed in a plane crash near Point Barrow, Alaska, on August 15, 1935. He did not get to witness the implementation of the shelterbelt policy. In that same year, the federal government launched its massive tree-planting campaign. In total, the government planted more than 218 million trees on 30,000 farms, forming 20,000 miles of windbreaks.[40]

Today, shelterbelts reduce soil erosion and wind desiccation on plants. Shelterbelts are planted on the western sides of farms or ranches in a north-south direction. Because these trees must be able to survive on less than thirty inches of precipitation a year, the variety of trees grown in a shelterbelt is limited. Red cedar, spruce, and pine are frequently grown along with rows of grains in between them for extra erosion control. Shelterbelts are made up of anywhere from one to five rows of trees and grains. Studies show that a thirty-mile-per-hour wind striking a five-row shelterbelt is reduced to eight mph on the immediate leeward side. At three hundred feet on the leeward side, the velocity is still only fourteen mph. Some protection against strong winds is measured at 1500 feet east of the shelterbelt.[41] Unfortunately, before the Roosevelt administration's efforts to stabilize farm conditions in the 1930s, shelterbelts were virtually unheard of among the farm and ranch populations of the western Great Plains.

In addition to the shelterbelt policy, Roosevelt established the Soil Erosion Service (SES) in 1934. The SES conducted forty-one soil and water conservation test projects and the newly established Civilian Conservation Corps (CCC) provided the labor for the schemes. The completion of these projects impressed Congress so much that it created the Soil Conservation Service (SCS). The Rural Electrification Administration (REA) was established in 1935 with the intent to help electrify farms, which had heretofore relied on wind power. By 1950, 78 percent of farms and ranches in the United States were able to replace windmills and wind power with electrical power generated in the Tennessee River Valley and delivered to the Great Plains through cooperatives.[42] This innovation enabled farmers and ranchers to draft larger volumes of water from the Ogallala Aquifer.

The Ogallala is a sandstone aquifer made up of materials eroded from the Rocky Mountains. The porous rock formation stretches through the Texas Panhandle and eastern Colorado to western Oklahoma and western Kansas. It then spreads eastward and underlies roughly four-fifths of Nebraska. It reaches a maximum thickness of nine hundred feet in southwestern Ochiltree County, Texas.[43] In total, the Ogallala lies beneath 225,000 square miles of high plains. Besides providing water for irrigation, it also furnishes drinking water for over two million people. As a consequence of depleting groundwater faster than nature can recharge it, the costs of pumping increasingly deeper water to the surface and then to storage containers is adding to regional economic woes. At some point in the near future, the aquifer will be completely depleted.

Exacerbating the situation, according to Donald Wilhite, is global warming.[44] He contends that the western High Plains region, under the influence of global warming, will become drier, and as a result, soil moisture as well as atmospheric conditions will become more arid. Unfortunately, he sees the drying of soil and the depletion of the Ogallala Aquifer as a consequence of global warming when in fact it may actually contribute to it. While a great deal of attention is given to the role of carbon dioxide in climate change, humidity is an equally culpable heat-trapping element. The addition of 650 trillion gallons of ancient Ogallala groundwater into the hydrologic cycle, which has water changing forms through evaporation,

Image 5.4. An aerial view of the western High Plains. *Image courtesy of the author.*

condensation, and precipitation, is increasing humidity and may well be contributing to global warming. Most of the water mined from the Ogallala has been in a subterranean liquid state for millions of years. Since many municipalities around the world also rely on groundwater to support their populations, it begs the question: Is there a better way to irrigate with water already present in the hydrologic cycle?

SUMMARY

In this chapter, we looked at air pollution issues, the death and rebirth of Lake Erie, and problems associated with the Dust Bowl and subsequent groundwater depletion in the North American Great Plains. While environmental determinists likely view these situations as evidence that the environment had indeed sent clear signals to stop our exploitative activities, possiblists argue that human ingenuity has allowed us to develop and implement technologies that permitted us to overcome heretofore environmental limits. From a human ecology perspective, it must be recognized that in many respects both views have merits. Beyond them, however, is the awareness that nature, through its powerful forces, provides us with systems that beg to be understood and harnessed for our own well-being. Constant diligence is required to ensure the sustainably of Lake Erie; in the case of the Great Plains, groundwater depletion needs to be replaced with technologies that harness the hydrologic cycle. Outside of these ecological issues, there are shifting economic patterns around the world that have perhaps provided Westerners with a false sense of accomplishment that their policies aimed at reigning in spillover costs associated with industrial pollution were successful in improving air and water quality in places like Donora, Pennsylvania, and Lake Erie. When industrial plants shut down, clearly their wastes will no longer be a factor in the area's environmental health. Such is the case in the American Rust Belt. As environmental laws were passed in the United States during the 1960s and 1970s, manufacturing investments in Asian countries increased. Although the air above New York City and Pittsburgh is much cleaner today than it was in the 1950s, factories in ASEAN countries and China are churning out low-cost products at a mind-boggling rate. Their productivity is a by-product of fabricating natural resources in societies with few environmental safeguards and abundant, low-cost labor supplies. While it is beyond the scope of this book, it is worth pointing out that North America (Panama to Canada) has about 460 million people. Household incomes among residents vary widely, so the consumption rate among them is not uniform. By contrast, there are over three billion Asians living and working in societies that are rapidly industrializing. The heretofore industrial nations of the West and Japan once served as the primary markets for their products, but as disposable incomes among Asian countries rise, the UN and relatively developed countries of the recent past have little outside of logic to influence them to safeguard the environment. Asians are now marketing to Asians.

As Asian incomes rise and the continent's environment suffers under loose production safeguards, there will be increased pressure from within their own populations to change. There is no reason to believe that those countries will not adopt effective and perhaps even better environmental safeguards than were previously developed in the West. Some observers in the United States and Europe see this economic scenario developing and believe that their countries are well positioned to lead the way in green technology. Unfortunately, in the near future, countries that are producing cheap products will likely take "green" innovations researched and developed in the West and then manufacture them in their own countries. With current labor supplies as they are in Asia, Africa, and Latin America, it will be some time before the costs of labor and manufacturing around world stabilizes. Nevertheless, there is much that can be learned from the North American experience in coping with the forces of nature.

As we will see in the next chapter, dealing with water-related issues is not a problem reserved only for residents of the semiarid High Plains, including those who lived (and died) in the Dust Bowl. The southeastern United States has had to contend with times in which there is too much water. Moreover, the abundance of water in the region was used to produce electricity that enabled High Plains farmers and ranchers to pump irrigation water from the Ogallala.

Chapter 6

EROSION AND DISPOSSESSION IN THE CUMBERLAND GAP AREA, 1933–1939

INTRODUCTION

Even as Dust Bowl winds blew Oklahoma prairie soil onto ship decks in the Atlantic, it was argued that rainfall and streams carried precious southern Appalachian soil into creeks and rivers that eventually emptied into the Gulf of Mexico. While the Dust Bowl was certainly a geokolasis[1] crisis with clear impacts on people who were culpable in the calamity,[2] their situation was nevertheless humanely chronicled by John Steinbeck in *The Grapes of Wrath*.[3] Whereas nature forced thousands from their farms in Oklahoma, Texas, and Kansas, the building of Norris Dam as part of the Tennessee Valley Authority's (TVA) mission to reduce the deleterious impacts of erosion while generating electrical power and improving farm production resulted in the permanent removal of 153,000 acres of bottom lands from agricultural development and the dispossession of three thousand mostly subsistence farm families.[4] Because of the role played by erosion in the creation of Norris Dam and the TVA, it is inviting to see environmental factors as the catalyst for H. C. Darby's geography-behind-history approach to the study of the Cumberland Gap area in the New Deal era.[5] This chapter challenges that perspective and, at the same time, proposes a history behind geography delineation of the region during that critical time in American history.

Many historians and geographers, but certainly not all, have declared that the TVA, which was one of FDR's Hundred Days programs, was indeed "one of the New Deal's most popular and enduring achievements."[6] Historians Nelson Lichtenstein, Susan Strasser, and Roy Rosenzweig write that the "TVA was the New Deal's most ambitious and celebrated experiment in regional economic planning.... The authority soon tamed the flood prone rivers of the Tennessee Valley and, in the process, became the largest producer of electric power in the United States."[7] Environmental conservationists like Daniel Chiras and Oliver S. Owen also regard the TVA as a success and a model for regional development in other countries, including India. As they write:

> The establishment of the Tennessee Valley Authority (TVA) in 1933 was a bold experiment, unique in conservation history, to integrate the use of the resources (water, soil, forests, wildlife) of an entire river basin. Although highly controversial at the time, it has received international acclaim and has served as a model for similar projects in India and other countries.[8]

While conservationists, historians, and geographers have recognized the achievements of the TVA, the plight of the dispossessed was more than adequately described by Michael J. McDonald and John Muldowny.[9] Other writers, like Donald Davidson and Susan V. Donaldson have celebrated the stance that Southern agrarians took against modernization, and certainly the TVA and the Progressive Era epitomized that evolving social condition.[10] Because of the informality of the pre-1930s agricultural economy in the Cumberland Gap area, it is difficult to quantify the costs in loss of agricultural production to the region. Dispossessed farm families had little hope of finding jobs because the country was struggling with double-digit unemployment rates that reached 25 percent in 1933—the same year that the TVA was formed. According to Les Williamson, whose family lived south of the Cumberland Gap in the Powell Valley of Tennessee, a few local men were employed to build Norris Dam, but when the construction ended in 1936, so did their jobs.[11] This situation was especially difficult for many of the dispossessed because, like their nondispossessed neighbors, they had operated in the blurred world that existed along the threshold of subsistence and commercial agriculture.[12] While most dispossessed families owned the lands on which their homes sat, 6.2 percent were "renters," 4.5

percent were "share tenants," and 10.9 percent were "sharecroppers."[13] One out of five families was hence removed without compensation. As far as the landowners were concerned, John Rice Irwin, the founder of the Museum of Appalachia in Anderson County, Tennessee, recalled that he did not detect bitterness toward the TVA, except in regard to the paltry money people were paid for their lands.[14]

Most residents of the Norris Basin were members of faith communities. In the cemetery grounds adjacent to most churches laid the bodies of family and loved ones. Given the fact that this area of Tennessee is situated near the Cumberland Gap and Daniel Boone's Wilderness Road, which was blazed in 1775, some churches had witnessed the birth and death of generations dating back to the late eighteenth century. As the resting place for those early settlers as well as the recently deceased, church graveyards were regarded as sacred places. Constructing Norris Dam required the removal of an estimated six thousand graves.[15] It was only after public outrage that the TVA decided to allow churches and families to reinter the bodies in locations of their choosing. Officials had planned to reinter them in a national cemetery.

A point that is often missed by historians and geographers who write about the TVA is that the initiative was one of many of the Roosevelt administration's first Hundred Days programs. The Agricultural Adjustment Act (AAA) of 1933 was another of those early programs. Its primary purpose was to restrict the supply of food at the production end and destroy surpluses before they reached markets and drove down prices. The goal was to increase food prices, so commercial farmers' purchasing power would match that which they enjoyed in 1914. Commercial farmers in the South and Midwest were arguably among those who benefited the most from the AAA. Because a large portion of the residents in the Cumberland Gap area were subsistence farmers with some informal free-market activities built in to provide modest cash incomes, the region did not have the infrastructure to compete against established agricultural economies of the farm belt. It also meant that local farmers, let alone unemployed city dwellers, derived little benefit from the AAA. According to Rufus Voiles, an Anderson County resident who vividly recalled the 1920s and 1930s, "it didn't make any sense for the government to destroy all that food when there were hungry people lining up in soup kitchens."[16] With 153,000 acres of bottomland taken out of agricultural production, the formation of Norris Lake certainly fit the model of restricting food production to benefit commercial farmers.

Aside from the issue of food production, the impoundment impacted traffic flow. The reservoir became a physical barrier to residents in southwestern Claiborne and Union Counties. Before the reservoir was built, farmers living in Union County's Blue Springs and Braden's Chapel communities had travelled roughly fifteen miles to Maynardville. With the reservoir blocking the shortest route, residents were forced to travel nearly fifty miles to reach their county seat. The situation was somewhat better for Claiborne County residents, but in an age in which the best rural roads were graded and covered with gravel, even the shorter thirty mile trip to the county seat was cumbersome at best.

The construction of Norris Dam created a need for housing for TVA employees. Roosevelt and the TVA chairman Arthur Morgan decided to build an entire town to house the new agency's workforce. New homes and facilities were to model the benefits of electric heating and other modern conveniences. Also named for the Nebraskan senator, the original plans for the town of "Norris" included one thousand homes situated on one- to five-acre tracks. It was thought that residents could do a bit of farming to raise and sell produce at local markets or, at a minimum, to augment their diets. It was also hoped that these large lots would permit residents to open small businesses. However, the cost of building the community, which also included a school, dormitories, and paved roads, forced the project to be reduced to around three hundred housing units. While the town of Norris provided residences for some skilled workers from the region as well as for professionals from other parts of the country, there were no dispossessed families relocated to the model settlement.[17] To connect the isolated community to the city of Knoxville, a two-lane concrete road named Norris Freeway was built. In the wake of the building of the dam, the community suffered financially because of the exodus of construction workers. The town of Norris was henceforth regarded as a country retreat by the TVA's professional staff. For dispossessed locals, it was not much of an economic benefit.

It is clear that much is known about the plight of the dispossessed. For their neighbors who lived in more rugged places away from the Norris Basin, less is known; though some assumptions about them can be made. Given that erosion was a regional problem, it is reasonable to imagine that denuded landscapes and, hence, poor farm conditions prevailed and plagued the lives of the nondispossessed. However, more research on this topic is needed. The purpose of this chapter, therefore, is to cast light on the lives of Depression era families who remained on the land in the Cumberland Gap area. The research reported ahead will answer the following questions: Was subsistence farming the way of life? Did families have other sources of income, including jobs with the TVA or in the coal mining industry? Were families able to help others in need? Did families see themselves as better as or worse off than city dwellers? In short, were regional families well adapted to life in the hills and hollows of the Cumberland Gap area of Kentucky and Tennessee?

FRAMING A PERSPECTIVE ON LIFE IN THE CUMBERLAND GAP AREA

To answer those questions, inspiration is found in the notion of human adaptation, which is central to the widely embraced model known as human ecology. The interdisciplinary paradigm was originally advanced by Harlan Barrows, one of the Chicago School's leading lights. Human ecologists, including Barrows, argue that people adapt to the limits imposed on them by nature.[18] However, human ecology has attracted criticisms from at least one prominent historical geographer. According to renowned Cambridge University professor Alan R. H. Baker, Barrows's human ecology perspective was plagued by naïve analogies, crude empiricism, and functionalist inductivism.[19] In spite of Baker's limited appraisal of the paradigm, human ecology still garners respect from other geographers, some biologists like Paul and Anne Ehrlich, and some urban sociologists including Herbert Gans, Walter Firey, and William Schwab.[20] In response to some of the criticisms by the likes of Baker, human ecology has mutated into two subparadigms: the sociocultural approach and the neoorthodox approach.[21] While the sociocultural view emphasizes the role played by sentiments attached to family, political affiliation, and expressions of culture, ecological factors still influence people's choices in many public behaviors, including where they choose to live. For instance, a number of inner city areas across the United States have retained high property values while in other urban areas, inner city neighborhoods have deteriorated. The neoorthodox approach recognizes that problems plagued the paradigm's simplifications, especially those that argued that geographic communities were shaped by local environmental factors alone. Nonetheless, advocates of the neoorthodox approach still argue that technological innovations are examples of ways in which humans adapt to various environments. The paradigm that inspires this research embodies elements of the sociocultural and neoorthodox perspectives.

Two recent publications highlight the application of human ecology to Appalachia in the 1920s and 1930s, although one sees cultural change as the product of the macroforces of government while the other views adaptation as the result of local, community-based processes featuring a host of human-environmental interactions. Associate professor of history at Rutgers University Neil M. Maher delves deeply into human-environmental relationships that characterized life during the New Deal era in his *Nature's New Deal: The Civilian Conservation Corps and the Roots of the American Environmental Movement*.[22] Maher provides detailed accounts of the impact of the Civilian Conservation Corps (CCC) on poor people who found employment in the agency's efforts to plant trees, clear trails, build parks and campgrounds as well as help ensure ecological balance. While serving the greater good through policies and projects, in Maher's view, the TVA and the CCC set aside green spaces that facilitated outdoor recreational opportunities for future generations. His book also shows how Roosevelt used conservation as a catalyst to forge his version of state building.

Assuming a perspective focusing on the microspaces occupied by families and communities, environmental historian Sara Gregg at the University of Kansas describes Appalachian folk in the early twentieth century as self-sufficient. She argues that contemporaries widely disparage the subsistence farms that operated in the northern Blue Ridge region of Virginia. However, from her perspective, these farms were ecologically viable, strong, and self-sufficient.[23] While her research has relevance to the present study, her study area of the northern Blue Ridge is more than two hundred miles from the Cumberland Gap area. In some respects, this book extends Gregg's research by adding first-hand accounts of residents who lived in seven counties in the Cumberland Gap area.[24]

As suggested by comments to me made by residents in this area of the South, theirs was a self-sustaining *Gemeinschaft* world in which localism prevailed. The informal, community-based life in the area meant that official data on economic output in the agrarian economy are suspect and certainly conservative, so any comparison to the impact that the TVA may have had on money transactions or quality of life in the agricultural economy are virtually meaningless. A qualitative approach illuminating the living conditions of the region's residents in the 1930s is needed. Such is the purpose of this chapter. To that end, potential informants in the area were identified through regional college and public school personnel.[25] Sixty-three people were asked to participate; fourteen took part in the study, including nine women and five men.

TENNESSEE VALLEY AUTHORITY AND NORRIS DAM

A brief review of the history of why the confluence of the Powell and Clinch Rivers was chosen shows that politics and a desire to provide electricity to cities and commercial farmers elsewhere may have been the primary motives for building Norris Dam. Indeed, President Roosevelt said as much:

> Before I came to Washington [in 1932], I had decided that for many
> reasons the Tennessee Valley—in other words, all of the watershed of
> the Tennessee River and its tributaries—would provide an ideal land
> use experiment on a regional scale embracing many states.[26]

However, the idea of using the Tennessee Valley as a land use experiment did not originate with FDR. In 1916, President Woodrow Wilson signed the National Defense Act. Funds allocated by the act enabled the federal government to purchase a large tract of land for processing nitrates on the Tennessee River at Muscle Shoals, Alabama.[27] The aim was to break the country's dependence on importing nitrates from Chile. Nitrates are an active ingredient in fertilizers and conventional explosives. To provide power for nitrate production, electricity was generated at Muscle Shoals in a newly constructed impoundment that was handily named Wilson Dam.

Private industry and members of Congress in the 1920s saw great potential for using water power to generate electricity, but few places in the United States offered the geographic features necessary for moving ahead with the technology. The question of what to do with the war-era dam built at Muscle Shoals was hence a debated subject. Some politicians argued that it would be best to place the power-generating structure in the hands of private owners while others, including Senator George W. Norris (R-NE) felt that power generation was too important to the nation to be subjected to capricious market forces. The public, he argued, needed reliable and relatively cheap power, so its generation should be controlled by the government. Wilson Dam did not produce enough power for places located beyond those surrounding Muscle Shoals, so more dams were needed. Wilson Dam happened to be located in the southern end of Appalachia and the lowest elevation point in the Tennessee Valley in the eastern section of the watershed. In the minds of Progressive politicians and possiblists, the tributaries that drain into the Tennessee River now had potential for the greater and collective good.

Southern Appalachia was ideal for Norris's goal to generate electricity for rural America. As a Progressive, Norris had long contended that rural electrification would help drought-stricken farmers and ranchers in the farm belt. From a civil engineering perspective, southern Appalachia was ideal for producing hydroelectricity. Unlike the rugged and wet northeastern parts of the country, the region seldom has cold enough temperatures to freeze rivers and streams. The valley has abundant and dependable rainfall with most of it falling in winter and spring months, when lower evaporation rates allow for higher volumes of ground and surface water flow. The region's ridge and valley topography could serve as conduits and retaining walls to impound water in front of a concrete and steel-framed dam. The area had one more important resource: patriotic, cheap, and readily available labor. All that was needed to bring Norris's vision of cheap power to his beloved heartland was national will and a like-minded president. Presidents Warren Harding, Calvin Coolidge, and Herbert Hoover, who occupied the White House during the 1920s, were mostly adherents of free-market capitalism, so they were resistant to such a huge role for the federal government to play. The Roaring Twenties was a time of great optimism and was highly compatible with the geographic paradigm known as possiblism—but it ended with the dismal collapse of the Stock Market in October 1929, thus setting the stage for a perfect storm of political, economic, and environmental crises that changed the national will. Unfortunately, the highly coveted watershed was well populated. The Cumberland Gap area had been settled by whites since Colonial times, and residents had steadfastly, although not uniformly, been loyal to the United States during the Civil War.

With rumors of war emanating from the early rumblings of the second world-wide conflict fueling a sense of nationalism among residents, they believed that they rather spontaneously participated in the TVA for the national and perhaps even divine good.[28] However, the fact that the TVA was a Hundred Days project suggests that powerful people had already decided on ways to use the valley before Franklin D. Roosevelt was elected. As FDR described the genesis of his plan: "This plan fitted in well with the splendid fight which Senator Norris had been making for . . . the Wilson Dam properties. . . . Senator Norris and I undertook to include a multitude of human activities and physical developments."[29] These plans were in place at least as early as 1931, some two years before Adolf Hitler rose to power in Germany and a full decade before the United States entered into WWII.

Because water in the Tennessee Valley flows in a southwestern direction toward Muscle Shoals, the first major dam initiative had to be built at a higher elevation in the water shed. Such a place was identified at the confluence of the Powell and Clinch Rivers near the Cumberland Gap. As the valley area with the highest elevation and suitable for dam construction, the land around the confluence was renamed (without local input) the Norris Basin; Norris Dam became the first impoundment built by the TVA. It was completed in 1936. Eighteen more dams were built by the agency before Roosevelt's death in 1945.

Getting electricity from the eighteen dams in the Tennessee Valley to commercial farmers elsewhere required another engineering feat; therefore, a series of electrical cooperatives were built by funds made available through the Rural Electrification Act of 1936. Senator Norris and Mississippi representative John E. Rankin cosponsored the bill. Electricity generated by the TVA was used to light farmhouses and run pumps to mine groundwater for irrigating corn and cotton fields in the Deep South and the Midwest. Later in Mississippi, electricity was used to supply water and oxygenation equipment for catfish farms. This situation begs the question: For whom was this aspect of the New Deal intended?

While most writers rightly give "credit" to Roosevelt for making possible the environmental policies of the New Deal, there were other key players who worked tirelessly to bring about cultural and environmental change in southern Appalachia. The Tennessee Valley Authority's first board of directors was made up of well-educated men from other parts of the country and Canada. At least one of them harbored attitudes of ethnic superiority. One such man, named Arthur Morgan, was the president's handpicked choice to serve as chair of the board. The board member with the closest connection to the region was Harcourt Morgan (no relation to Arthur Morgan), although he was a transplanted Canadian by way of Louisiana. In 1933, he was the president of the University of Tennessee. Morgan had risen from the ranks of the faculty of agriculture at Louisiana State University to serve as director of the University of Tennessee's experimental station. He then served as dean of agriculture before assuming the presidency. He had a reputation for getting along well with faculty and students and was regarded as a friend to farmers. Chairman Arthur Morgan, on the other hand, was raised in Minnesota and had served as president of Antioch College in Ohio. He was trained as an engineer, and his biographer, Roy Talbert, actually described him as a human engineer: one who

could use technology to fix apparent social and cultural problems. It is
interesting that Talbert characterized his mind as being affected by late-
nineteenth-century race theory that featured Anglo-Saxonism and
eugenics.[30] Because Appalachian folk were mostly of Celtic extraction
(Irish, Scottish, Welsh), Arthur Morgan most likely held quite negative
views of them. After appointing two directors, an engineer and an agriculturalist, to the new TVA board, the president felt that the third director
needed knowledge of utilities. He turned to David Lilienthal, a thirty-
three-year-old who had served as a member of the Wisconsin Public Service
Commission.[31] With the exception of the Canadian Morgan who had at
least lived in the region, FDR, Norris, and the other members of the TVA
board were no doubt operating on assumptions about the Appalachian
people and their lands, and in the 1930s, those images were not positive.
Indeed, locals were regarded as deficient. However, in the next section of
this chapter, attention will be given to how nondispossessed residents of the
area lived. Wayne Flint, a retired Auburn University professor of history,
described long-held views of poor people in the South, including the
southern reaches of Appalachia. "Poor whites have been characterized as
degenerate racists, white trash commonly guilty of incest and mindless violence. Primitive and illiterate, they supposedly bequeathed no legacy to
music, literature, art, or architecture. Their society was clannish and closed
to outsiders, who were objects of fear and suspicion. Economically, they
were impoverished people who lived a marginal existence."[32] With the
exception of Les Williamson mentioned previously, none of the interview-
ees mentioned the TVA in answering the researcher's questions (see
appendix for the survey questionnaire).

LIFE OF THE NONDISPOSSESSED

As observed by Verna Mae Slone in her book *What My Heart Wants to
Tell*,[33] an unknown number of Appalachian folk have suffered a loss of personal and communal pride by accepting negative stereotypes and hurtful
monikers.[34] In the 1930s, a European visitor made an observation of life in

Appalachia that seems to epitomize the negative and backward view of the people and their culture:

> [A] very large percentage of them had kitchens with ovens burning wood. . . . They were lighted by dim, smoking, smelly oil lamps. . . . The washing of clothes was done by hand in antiquated tubs. . . . The water was brought into the house by women and children, from wells invariably situated at inconvenient and tiring distances. . . . Ordinarily there is no icebox, so many products that might be grown to vary the horribly monotonous diet are out of question; they could not be stored.[35]

The poignant story of impoverished Southerners, including those in the Appalachian region, is well chronicled by notable writers such as Slone, Ronald D. Eller, Wayne J. Flynt, Harry Caudill, Henry D. Shapiro, and me, among others.[36] In the 1930s and even later, the outside world saw those who lived in the hills, "hollers," and hilltops of northern Alabama, northern Georgia, eastern and middle Tennessee, western North Carolina, northwestern South Carolina, eastern Kentucky, southwestern Virginia, and all of West Virginia as either unlucky Anglo-Saxons or the descendants of pickpockets and thieves and people who otherwise were genetically predisposed to poverty, incest, and backwardness.[37] While those seminal works are invaluable in shedding light on life in the southern mountains, they do not specifically address living conditions in adjacent areas targeted by the TVA for building reservoirs. As mentioned earlier, authors McDonald and Muldowny wrote about the plight of the dispossessed, but their general description of living conditions likely applies to those who lived in areas away from the reservoir. "The inhabitants of the Norris Basin, the first host population of [the] TVA, were for the most part isolated and economically disadvantaged, owing largely to geographical location and to the pressure of population on marginally productive lands."[38] It is generally thought that this situation placed more stress on the land and reduced the quality of life found in the valley. According to Les Williamson, whose maternal family has lived in the Cumberland Gap area since the late 1700s, inheritance was a major culprit in reducing his family's ability to stay on the land: "My family's once huge landholdings in the Powell Valley was divided up among heirs for too many generations. By the early 1930s, my paternal grandparents' parcel was too small to farm, so they moved to nearby Middlesboro where they unsuccessfully operated a bus line and a mattress company."[39] A similar "overpopulation" situation befell the Charles Lane family of Powell Valley.[40] There is hence some credence to the observations made by McDonald and Muldowny.

The shrinking of family farms like that of Les Williamson's family in Powell Valley was not a concern of the federal government, because restricting agricultural production was included among the programs of the famous Hundred Days of the New Deal. The Agricultural Adjustment Administration (AAA) established by Congress in May 1933 was intended to limit the output of American farms. Commercial farmers were paid to plow under crops (running a plow over various crops, thus destroying the produce on the ground and then using the plant material as organic fertilizer), including ten million acres of southern cotton. Also, hogs were slaughtered; nine million pounds of healthy pork were buried.[41] Because millions of Americans were unemployed and soup kitchens were springing up in Rust Belt cities, these actions produced a good deal of public outrage. Nevertheless, destroying food was done in the name of adjusting market forces to raise farmers' purchasing power to its previous, 1914 level. However, because of the informal nature of the way in which goods were traded, this policy had little relevance to residents of the Cumberland Gap area.

While most farm residents had little cash, bartering of surplus produce took place, but charity was perhaps just as common. Linda Keck grew up in the Speedwell area of Powell Valley. She described her parents' role in bartering labor for goods: "Both sides of my family farmed but there were times [when] my mother's dad walked across the mountain about two and a half miles, one-way and cut firewood to be sold. He would be paid with flour or other types of dry goods. My mother recalls saving the flour sacks, washing them, and making dresses. A twenty-five-pound bag of flour sack was one square yard of fabric."[42]

Vernedith Voiles, whose relatives were removed from their farm by the TVA, recounted instances in which travelling "city folk," often on foot, stopped by their "subsistence" farm to ask for a meal in exchange for work.[43] Although her family lived just north of the Cumberland Gap area in Rockcastle County, Kentucky, Mary Ruth Isaacs nonetheless describes similar situations in which her grandparents helped those in need of food: "My grandparents did not have any real wealth, but they never turned anyone away who was hungry. My grandma said that 'tramps' would come along from time to time, and she would be afraid to let them in the house. She would tell them to sit outside and then she would bring them a plate of food. No one was ever turned away without eating."[44]

Les Williamson stated that a railroad line ran nearby his maternal grandparents' home in the Arthur area of Powell Valley, which sits about four miles to the south of the Cumberland Gap. "It was not uncommon for hobos from some distant place to knock on the door looking for food. They weren't refused."[45] Terri Keeton, whose family continues to live on the Cumberland Plateau that forms the western wall of the Powell Valley, also described her grandmother's home as a place where hobos off of nearby trains would ask for food. Her grandmother "would give them her last bite to eat, for she felt it was her responsibility, as a Christian, to feed those less fortunate."[46] Dorothy Jones stated that her family did not have much money, but they "were much better off then [sic] city people."[47] Despite having a home and food to eat in a time of national dearth, negative observations of the culture suggested that local residents lived in miserable conditions characterized by "no iceboxes" and monotonous diets. In reality, Vernedith Voiles claimed that most homes had springhouses in which milk, butter, and other perishables were kept cool for at least a few days. Meats were smoked and/or salted to preserve them, and canning enabled mountain people to eat a variety of vegetables throughout the winter months. Jimmie Edwards (born in 1935) grew up in Union County near the newly formed Norris Lake. As a child in the 1930s, she recalled:

> Momma canned everything she could get her hands on. We had a well and cistern, grew tomatoes, green beans, cucumbers, and raised hogs and chickens. Beef was rare at our house. Daddy and my brothers brought in lots of squirrels and rabbits. We never went hungry. Our

sweet tooth was satisfied by all the wild strawberries, blackberries, and mulberries that grew on the hillsides among the thickets. We'd go out and pick two gallons of them, and momma would can some and [the] others she would use to make a pie or dumplings.[48]

Rosemary Weddington, who was born in 1930 in Bell County, Kentucky, stated that her family lived on a farm that was located near Varilla, a once prosperous coal camp:

[We had] a farm with cows, chickens, pigs, [and the like,] and planted about an acre and a half of ground with vegetables. My father built a cellar where potatoes, apples, and canned goods were stored. My mother did a great deal of canning and all kinds of goodies were stored in the cellar and kitchen cupboard. We were never without food. Mother even baked bread and goodies. I didn't really know there was a depression. My later childhood was a happy time and we (at least I) didn't know about the Depression. I would hear my parents and kinfolks talking but I was still a happy child.[49]

However, Ms. Weddington's home was near a coal camp that was economically tied to industrial production up north. Some locals lost their jobs in the mines, and family members who had gone north to work in factories returned to the region where they were helped by relatives. Weddington further recalled:

[I] can remember many times various people would knock on our back door porch and ask for work or a handout. My parents were good Christian people and could never turn them away. They usually fed them, gave them food for their next meal, and even gave them clothes and shoes. If my father needed them to work, he would pay them with food, chickens, [and the like]. My mother always told me after these people left that "but for the grace of God, there go I." And she would add, "Always be grateful for what you have and share it with those in need." [This was] probably my best lesson in childhood.[50]

Indeed, observations of unemployed people in coal camps are further illustrated by the comments of Oline Carmical who was born and raised in Harlan County, Kentucky:

> My father was an underground coal miner who was often out of work due to labor strikes and the more than usual economic downturns of the era. My mother was a first grade teacher in the public schools. Before they married on January 2, 1937, and thereafter, they lived in the Clover Fork [Evarts] section of Harlan County, Kentucky. There is and has been little farming there.[51]

Life in coal mining areas was not as bad as those experienced by families like the Carmicals if they owned land with tillable soil. Sue Wake, who was also from Harlan County, claimed that in comparison to city residents and a coal miner without land, her family was "better off because they had a garden, and they had jobs in the coal mines."[52] The ability to farm and mine coal, as suggested by Sue Wake, indeed made life somewhat better for those who could. Diana Baker's family lived in Letcher County, Kentucky, which borders the Powell Valley of extreme southwestern Virginia. As she recalled:

> Some family members dug coal for coal companies and others sold the coal available to them on their own land. Corn grown on the hillside farm was used to make moonshine. The moonshine was sold in a dry county. Transporting moonshine was one way young men made some money. Grandmother grew chickens and traded eggs for bacon [obviously an example of the informal bartering nature of the agricultural economy]. She walked from Letcher to Knott County to trade for goods she needed. She used feed sacks to make quilts and some clothing. Teenagers went to the feed store and killed rats for a few cents each. Dead rodents were given to the store owner for currency. The rodents were not shot with a gun because ammunition was too expensive.[53]

Harold Hubbard was born and raised in Clay County, Kentucky. In addition to farming, his parents ran a grocery store and a post office. His grandfather ran a lumber mill and a mill for grinding corn. When asked if unemployed people sought help from his family, he said, "No, the Depression did not affect the Clay area, because people were very self-sufficient."[54] As noted earlier, Sara Gregg made a similar observation of subsistence farmers in the northern reaches of the Blue Ridge Mountains.[55]

The rural Appalachian culture may well have been considered premodern by urban elites, but it was vibrant enough for unemployed city dwellers and coal miners to find temporary work and sustenance on its Depression-era farms.

LINGERING LEGACY OF THE TVA IN THE CUMBERLAND GAP AREA

In 1933, places characterized by steep slopes were seldom chosen for home sites. The soil was too shallow for deep plow farming, so cattle grazing and timber harvesting were intensively practiced. In an agrarian society like the one dispossessed by the TVA in building Norris Dam, those who lived on hilltops and the sides of mountains were considered poor. It was inviting for others, especially those who farmed "bottom land," to attach negative monikers like "ridge runner," "hilltopper" and "hillbilly" to those families.

Nowadays, formerly marginal lands along the Norris Basin are attractive places for home sites. Lakeshore areas are especially attractive for those interested in recreational home sites. Close proximity to the lake also affects property values, which, in turn, adversely affects local residents. When available on the market, steep slope Norris Lake lots bring as much as $30,000 per half-acre. A five-acre home site with a lake view can sell for as much as $150,000. A similar-sized, moderately sloped lake lot can sell for as high as $5 million.[56] Lands located away from the lake, in the hollows of the Basin, sell for $5,000 to $10,000 per acre. According to Nancy Maiden, a local real estate broker, there are a number of famous celebrities and professional athletes who have built homes along the shores of Norris Lake.

Arnold Schwarzenegger reputedly built a cabin near the Deerfield community in Campbell County.[57] Maiden also estimates that approximately 70 percent of the homes built on the lake serve as recreational or second homes for nonresidents.

This evolving housing situation makes it difficult for local ways of life to continue on two fronts: first, with no thriving industrial Northeast to draw away low-skill workers, children of farmers are often given portions of their parents' lands because they cannot afford to buy land themselves. With smaller parcels of property, commercial agriculture will not be an option for succeeding generations of residents. Second, with nearly one out of four people in Campbell and Claiborne Counties living below the poverty level,[58] many dairy farmers have decided to sell off portions of their lands to nonresidents. In January 2010, for instance, a 17.42 acre track of pasture land without a lake view was on the market for $299,000.[59] With prices like this paid by nonresidents for pasture lands, it is no small wonder why farms are shutting down. According to Charles Lane, a former dairy farmer in the Powell Valley, he worked sixty to seventy hours a week and seldom made a profit. Lane's wife provided the bulk of the family's $25,000-per-year income by working as a clerk at a regional hospital.[60] Sadly, financial stress contributed to Lane's decision to close down the family farm. He and his wife also divorced, and with few employment options available to him, he relocated some two hundred miles away to Johnson County, Kentucky, where he works as a night watchman for the local high school. It is also revealing that, until he left the Powell Valley, he and only three of his ten siblings were able to make a living in the Cumberland Gap area. As young adults, the other siblings went to Ohio and Detroit, where they found work in automobile plants. Similar situations are found throughout the TVA-controlled watershed. The Lane family's plight is not isolated. Indeed, with productive pasturage taken out of agriculture, where will food be grown for an increasingly urban and suburban population? Food commodities will be controlled by fewer and fewer people, and instead more and more by large corporations that will control prices and availability.

Like the model town of Norris discussed previously, Norris Basin home sites have not become residences for the children of dispossessed locals;

rather, they have become the secondary and vacation abodes of relatively wealthy people from other places, like Rust Belt states and the West Coast. Clearly technical inputs have increased the market value of lands. In the Cumberland Gap area, it is increasingly difficult for families with modest incomes to buy land and build homes.

SUMMARY

Erosion and dispossession are odd but complimentary topics that acted together in the 1930s to affect an entire region on behalf of the rest of the country. In the case of the Cumberland Gap area, adherents of possiblism and political Progressives saw an opportunity to use flooding and erosion as the basis for taking hundreds of thousands of the region's farmlands out of production.

There is no doubt that Norris Dam has helped control flooding in the Tennessee River watershed. In the winter, when power generation is most needed and rainfall is plentiful, water passes through turbines creating electricity that heats and lights the homes of millions of people. Reservoirs retain less water during colder months, but as spring arrives and temperatures warm, less power is needed. Tennessee Valley Authority reservoirs are then allowed to fill to greater depths for the drier summer months. This annualized plan certainly eases flood pressure while providing recreational opportunities for boaters, campers, and others who enjoy fishing, or just want to cast a tranquil eye across placid waters. Bucolic images aside, the lake sits on top of fertile bottom land. Before the TVA, periodic flooding had allowed sediments to enrich the soil, but that cycle is lost along with access to those productive lands.

The establishment of the TVA coincided with the passage of the Agricultural Adjustment Act in 1933. Soon afterward, Congress passed the Rural Electrification Act (1936). Besides the president, the other key player in the passage of all of those new laws was George W. Norris. As a Progressive senator from Nebraska, he clearly wanted to help his constituents. With the western High Plains out of crop production due to the Dust Bowl and the Cumberland Gap area's potential to develop commercial agriculture halted by the TVA's land acquisitions, Nebraska's farmers stood to make the most out of others' environmental crises.

Despite losing 153,000 acres of farmlands, Cumberland Gap–area families were able to absorb surplus populations from urban places and coal camps that had been devastated by the Depression. While lifestyles were considered premodern by outsiders, comments made by residents to the author in interviews suggest that most of them were not ill fed. Indeed, through the use of canning vegetables and the salting and smoking of meats, food items were available throughout the year. Dairy products and other perishables were kept cool in springhouses. Though these homes were not powered by electricity, they were highly functional. Through further comments, it seems that intermittent work in coal mines offered more opportunities for non-farm-related work than the construction of Norris Dam (1933–1936). Meanwhile, some family farms had shrunk through inheritance to the point that they were too small for commercial agriculture. Nevertheless, for most of the nondispossessed, their lifestyles were well adapted to local environmental conditions.

Chapter 7

LIFE AND DEATH
IN HURRICANE ALLEY

INTRODUCTION

In the hours just past midnight on August 26, 2005, a hurricane with a well-defined eye and increasing strength was drawn over the steaming waters of the Gulf of Mexico. In its brief passage over southern Florida, Hurricane Katrina killed fourteen people.[1] By August 28, the storm had strengthened to category 5, the strongest rating for a hurricane on the Saffir-Simpson Hurricane Scale.[2] At its peak, the storm produced maximum sustained winds of 175 mph. When it struck southern Louisiana near Buras in the predawn hours of the twenty-ninth, it had lost some of its power but was still a formidable category 3 storm with constant winds of 125 mph. Katrina caused tides to rise or surge to ten to twenty feet above normal levels. The raging tempest then dropped eight to twelve inches of rain over the southern Gulf states before it weakened into an extratropical low over the Tennessee Valley and was absorbed into a westerly front.[3] In total, Katrina killed some twelve hundred people with most of the deaths occurring in Louisiana.[4] However, the death toll would have been much lower had there been a more efficient evacuation strategy in place. Sadly, many of those who died had likely heard the highly publicized warnings, but for unknown reasons, they did not heed them and leave on their own. It was the costliest storm in American history. According to Richard D. Knabb, Jamie R. Rhome, and Daniel P. Brown of the National Hurricane Center, the storm exacted an $81 billion toll on the United States.[5] Beyond those costs, a hurricane's impact on disaster insurance premiums is felt by homeowners throughout the country. Hurricane Katrina was indeed a

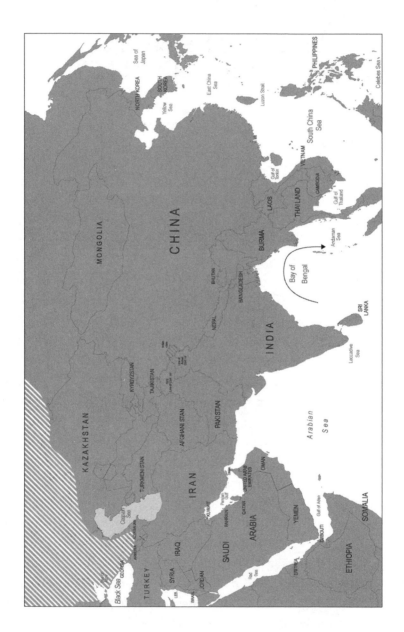

Map 7.1. Hurricane Alley: Bay of Bengal. *Image courtesy of Geraldine Allen and Clayton Andrew Long.*

national disaster of the first order, but it was by no means the country's most costly in terms of human life.

Even the storm that caused the worst loss of life in American history pales in comparison to the fatalities suffered by the countries surrounding the Bay of Bengal: Bangladesh, India, and Myanmar. While America's deadliest hurricane killed eight thousand people on Galveston Island on September 8, 1900, more than one-quarter million people lost their lives in a single cyclone (hurricanes are called "cyclones" in the Indian Ocean) that struck Bangladesh on November 13, 1970.[6] Twenty-one years later, another hurricane killed 138,000 Bangladeshis.[7] On October 29, 1999, nearly 10,000 Indians died in a cyclone.[8] Across the Bay of Bengal on May 5, 2008, a killer cyclone named Nargis made landfall in Myanmar. In its wake, 78,000 people were dead, and as of June 16 of that year, another 58,000 people were still missing.[9] In each of these cyclones, storm surge and flooding were the primary causes of death. The countries situated along the shores of the Indian Ocean's Bay of Bengal are in one of the most precarious places on earth.

As is shown in this chapter, states located along the Gulf of Mexico are the most likely victims of hurricanes in the United States. Like Bangladesh in the Bay of Bengal, New Orleans is located on the northern shore of a bay. The levies surrounding New Orleans, which in the city's lowest places are several feet below sea level, were built to prevent inundation from a hurricane's most potent killing weapon: the storm surge. Despite being struck in a near frontal assault from a category 3 storm, the city's levies held out against the storm surge. However, the twelve inches of rain that dropped inland in the Mississippi River Valley caused Lake Pontchartrain and the Mississippi River to rise well above their flood stages. The heavy precipitation also caused the water table to rise under the basin on which New Orleans sits. The city's levies, or any earthen structures for that matter, stood little chance against flood waters flowing from in front, behind, and beneath them.

The human death toll resulting from hurricanes, cyclones, and typhoons is not likely to lessen in the near future, for, as this book shows, human population along the Gulf of Mexico and the Atlantic Seaboard of

North America has exploded since the Industrial Revolution in the mid-nineteenth century and the Medical Revolution one hundred years later. With our seemingly innate desire to live near estuaries and fertile shores, population is growing in coastal plains areas of the tropics and southern midlatitudes because more temperate places are well populated. Technology gives us the sense that we can overcome nature's forces in these warmer, seemingly benign places. In this evolving environmental dilemma, one finds tension between environmental determinists and possiblists. Is nature telling humans to stay away from these bountiful places, or do humans possess the knowledge to develop houses and other structures that can withstand 125 to 150 mph winds and the objects hurled by them as well as storm surge and groundwater flooding?

As a discipline, geography (a parent of climatology and meteorology) offers us a deeper understanding of the appeal and consequences of living in hurricane alley, which includes the eastern shores of North America, the Bay of Bengal, and the coastal waters of Southeast Asia, including the South China Sea. In an application of geography behind history, this chapter will look at the geography of hurricanes and then offer an analysis of the hurricane that nearly washed away Galveston, Texas, on September 8, 1900. Even in comparison to behemoth killers like Hurricane Andrew, the Galveston storm caused the worst loss of life from any force of nature in American history. This chapter will conclude with a discussion of the potentiality for more direct hits from tropical killers.

GEOGRAPHY OF HURRICANES

As suggested by the locations included in hurricane alley, these tropical monsters form and live out their short lives in relatively few places on the planet. While an individual hurricane's path may defy exact prediction—thus making living in hurricane alley even more precarious—taken as a collective natural phenomenon, they do follow narrow swaths over three of the earth's four oceans. Due to a lack of solar energy absorption, the Arctic Ocean does not experience hurricanes.

The use of the designation "hurricane" to describe a giant low-pressure center producing wind speeds exceeding seventy-four mph is commonly used in the Americas and Europe. In the Indian Ocean, they are referred to as cyclones, and in the eastern Pacific Ocean, they are called typhoons.[10] The National Oceanic and Atmospheric Administration (NOAA) employs the Saffir-Simpson hurricane wind scale, which is based on a 1–5 rating (see table 7.1). The scale only considers wind speed, although some writers insert barometric pressure and the size of storm surge into the mix. The scale does not take into account the amount of damage caused by storm surge, tornadoes that form in the eye wall, and rainfall-induced flood events.

Table 7.1. Saffir-Simpson Hurricane Wind Speed[11]

Category	Mean Wind Velocity (mph)
1 Weak	74–95
2 Moderate	96–110
3 Strong	111–130
4 Very Strong	131–155
5 Devastating	>155

Wind velocity and direction are produced by differences in air-pressure centers (gradients) and the Coriolis effect. Hurricanes are giant low-pressure centers. In the Northern Hemisphere, air rises in a counterclockwise motion.

While powerful storms often make landfall along the west coasts of the Atlantic and Pacific Oceans, the most devastating storms in terms of loss of life have occurred in the Gulf of Mexico and in the Bay of Bengal. Because of their restrictive shapes, water slowly circulates in these relatively small and shallow areas of the earth's oceans. Inviting more volatile conditions, countercirculating eddies often form under the influence of high-pressure centers. Semitrapped water in the Gulf and in the Bay of Bengal is then exposed to nearly direct radiation from the sun, so it has the capability of storing and then releasing massive amounts of heat into the atmosphere. The shorelines and inland areas in these two places are extensions of their gently sloped sea floors. Coastal areas gradually climb in elevation, but few places within one hundred miles of the shore are higher than two hundred

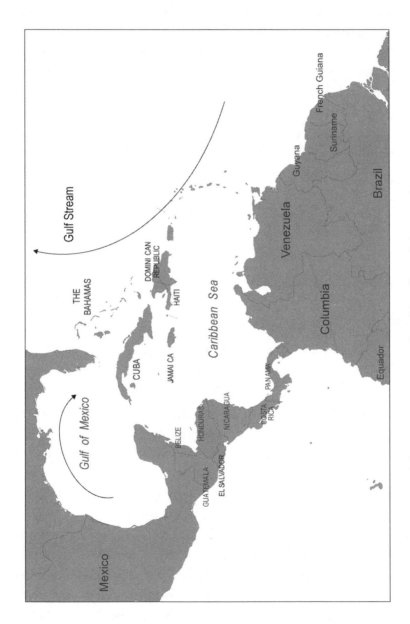

Map 7.2. Hurricane Alley: Gulf of Mexico. *Image courtesy of Geraldine Allen and Clayton Andrew Long.*

Image 7.1. A South Carolina duplex on stilts. *Image courtesy of the author.*

Image 7.2. California's Gold Bluffs Beach. *Image courtesy of the author.*

feet. Because of their gradually sloped landscapes, settlements along either sea coast are susceptible to storm surge.

The impact of storm surge is affected by the time of day that a hurricane makes landfall. If such a situation occurs during low tide, the damage from a ten-foot storm surge is not nearly as bad as one that strikes during high tide. Inland topography is also a factor in damage potential. If elevation climbs abruptly from the shoreline, as it does in Northern California—a place not in hurricane alley—destruction from storm surge would be virtually nonexistent. Unfortunately, shores and inland areas in the United States that are prone to suffer hurricanes are mostly low-lying coastal plains with little elevation change for miles inland. Cities in these places are in an unsafe situation. Being below sea level, residents in New Orleans are certainly in danger, no matter how much technology is available to them and how much time passes before the next "big one" hits. This observation comes into focus more when one considers that Katrina was a category 3 hurricane when it made landfall. What if a category 5 visits New Orleans during high tide?

Between 1900 and 2000, individual states along the Gulf were hit by hurricanes more often than states situated along the Atlantic coast.[12] These tropical monsters are products of convection generated by waters above 80°F. More importantly for those living in the Gulf region is the fact that the earth's spherical shape and its rotation cause many storms to turn right before reaching the coasts of Alabama, Mississippi, Louisiana, and Texas, as well as Mexico. Because of the Coriolis effect,[13] water currents set in motion by easterly winds moving over the warm waters off the coast of Africa will, under normal circumstances, continuously bend to the right, making a full circle around the Atlantic basin in the Northern Hemisphere. However, the Coriolis effect diminishes within a few hundred miles of the equator. Equatorial water nevertheless flows westward with a portion of it entering the Gulf of Mexico. In regard to the forces influencing the direction of the Gulf Stream, it is helpful to recall that the Kuroshio or Japan Current was likely used as a migration conduit for humans from Asia who settled as far away from their homelands as South America some ten to fifteen thousand years ago. In the case of hurricanes, water moving across the tropics as part of the north equatorial current heat up as sunlight penetrates the upper layers of the Atlantic. Sunlight is the source of a hurricane's power, and, as a result, a hurricane's lifespan can only occur over a few months of the year and certainly in only a few places on the earth's surface. While destructive storms sometimes strike the Pacific coastline of Oregon and Washington, for instance, they are not in the same meteorological league, so to speak, as hurricanes.

The reason why hurricane season in the Northern Hemisphere begins in June and ends in late October is because this is the time of year in which the Northern Hemisphere is most affected by sunlight. Hurricane season is not perfectly matched to spring and summer months. This is because it takes water longer periods of time to heat up or cool down, so there is a time lag in hurricane season of up to one month past the spring and fall equinoxes. This explains why destructive storms may form just after the fall equinox, when the lower latitudes of the Southern Hemisphere are receiving direct sunlight.

Over the one-hundred-year span beginning in 1900 and ending in 2000, sixty-five major hurricanes made landfall between Texas and Maine. Of that number, two made landfall in June and three came ashore in July. All five of those early summer storms came onto land along the Gulf Coast. Because of the smaller volume of water that was warmed as it flowed along the equator, the already warm water is then subjected to frequently occurring eddies in the Gulf of Mexico. Eddies permit the water to stay under intense sunlight for comparatively longer periods of time, so the Gulf of Mexico heats up quicker than the Atlantic Ocean. However, as the summer progresses and the Gulf Stream is warmed along the Atlantic sea coast, states there are just as likely to experience hurricane activity. In total, the number of hurricanes that made landfall in the United States during the twentieth century rose to sixteen in August, but the number of major hurricanes that did so in September more than doubled to thirty-six.[14] Even as late as August, Gulf Coast states were struck by hurricanes twice as often as those situated along the Atlantic Coast. However, in September, Atlantic Coast hurricanes were more common than those that made landfall along the Gulf Coast. The tally of October hurricanes dropped to eight; however, late or early season hurricanes can be highly destructive.[15] This was the case with Hurricane Grace when it combined with two other low pressure centers to form the "perfect storm" that sank the fishing vessel the *Andrea Gail* on October 28, 1991.[16] Fortunately for families living along the coast of New England, the full force of that storm was out at sea. Nevertheless, Hurricane Grace and other northern atmospheric monsters that form over the Gulf Stream are actually one way that nature disperses surplus energy absorbed in the tropics.

The fact that ocean currents provide northern locations with energy in the form of sensible heat is a natural phenomenon that has allowed people to live in places like Skara Brae in Scotland and Mount Sandel in Ireland. Without the diffusion of energy via ocean and atmospheric systems, living in places farther north than 40° north or farther south than 40° south would be too cold, as the biosphere in those zones radiate more energy back into space than they receive from the sun.[17] Similarly, life between 40° north and 40° south would be too hot without a means to disperse surplus energy, and, fortunately for us, nature provides the forces we need to survive in places that would otherwise be too cold or too hot for life to flourish, although as the latter portions of this chapter will show, energy dispersal is not always a safe process.[18] Hurricanes and lesser tropical low-pressure systems are simply means by which the earth transmits energy in the form of long wave radiation from zones of surplus (40°north to 40° south) to zones of deficit. Ocean currents likewise help disperse surplus energy toward the poles.

That statement does not mean, however, that the entire earth is ideal or suitable for human habitation. As we saw in earlier chapters, the mostly midlatitude ecumene that existed before the Industrial Revolution offers the optimum living conditions for people. Technology has allowed the carrying capacity in heretofore inhospitable places to expand and environmental stresses that make human life difficult seem safer and more inviting. In response to that perception, more and more people in the United States have moved into hurricane alley. With modern communication technologies, the larger society is able to witness hurricanes and is thus interested in knowing more about them. As was the case in the aftermath of Katrina, human causation and politics have made their way into discussions about hurricanes. It seems humans have always connected natural forces to the acts of people. Invariably this tendency invites the belief that God (or, indeed, gods) punishes people for their sins (geotheokolasis). Even nonreligious people sometimes believe that nature itself punishes environmentally insensitive people (geokolasis). This later point was made clear in the months post-Katrina.

After the flood waters were pumped out or had otherwise receded from the New Orleans basin, news stories were published suggesting that human-caused global warming was behind the increase in the number and intensity of hurricanes between 1995 and 2005.[19] According to *USA Today* writer Julie Snider, British scientist John Lawton and Germany's environmental minister Jurgen Trittin "have been quick to assert a connection."[20] The current cycle of more and deadlier storms could last fifteen to twenty more years, notes Snyder—although six years post-Katrina, this unsavory scenario has not happened. Despite noting the objections to claims of human causation in hurricane frequency by Max Mayfield, director of the National Hurricane Center, and meteorologist William Gray, Snider concluded her article by arguing that "it's worth researching whether global warming is affecting the frequency and intensity of those storms, but there's certainly no proof at the moment."[21] Gray contends that hurricanes Rita and Katrina, which struck only six weeks apart, mimicked the frequency of similar storms back in 1915. Given that the frequency and intensity of Atlantic hurricanes is somewhat correlated to patterns of global warming, few writers have argued that the relationship between hurricanes and global warming is just the opposite of the positions advanced by the likes of Trittin and Lawton. Such an argument would fit an environmental determinist paradigm, but it does not square with geokolasis, which seems to prevail in the minds of these prominent climate observers. For some reason, which is perhaps buried deep in the human psyche, many feel that humans are responsible for whatever befalls us, despite the logic of science and the laws of nature. While continuing the human blame mantra, a basic misunderstanding of how energy flows through the hydrologic cycle is seen in an article written by Jeffrey Kluger:

> Just why some areas of the world get hit harder than others at different
> times is impossible to say. Everything from random atmospheric fluc-

tuations to the periodic warming of the Pacific Ocean known as El Niño can be responsible. But even if all these variables have combined to keep the number of hurricanes worldwide about the same, the storms do appear to be more intense. One especially sobering study from the Massachusetts Institute of Technology found that hurricane wind speeds have increased about 50 percent in the past 50 years. And since warm oceans are such a critical ingredient in hurricane formation, anything that gets the water warming more could get the storms growing worse. Global warming, in theory at least, would be more than sufficient to do that. While the people of New Orleans may not see another hurricane for years, the next one they do see could make even Katrina look mild.[22]

Increasing global temperatures could indeed cause sea levels to rise, which is a process that allowed humans to settle in northern Europe, Asia, and the Americas some ten to fifteen thousand years ago. In truth, and in opposition to Kluger and proponents of geokolasis, sunlight, the energy source for heating oceans above 80°F, is the basic requirement for hurricanes to develop and then contribute to global warming via increased evaporation and condensation. Atmospheric moisture or humidity is a greenhouse gas. A hurricane's contribution to global warming happens because the hydrologic cycle serves as a medium to transmit energy. When heat causes water to evaporate, the process converts kinetic or working energy to latent or stored form, but when gaseous water (humidity) condenses, it releases the stored energy back into the atmosphere, creating more instability in the air. When atmospheric conditions are ideal (cooler air aloft), a hurricane can form. An average hurricane disperses about the same amount of energy into the atmosphere as electricity-producing facilities in the United States in a single year.[23] An investigation of long-term hurricane activity and global warming by Kenneth Trenberth points to the notion that warmer surface temperatures in the Atlantic and Pacific may be linked to increases in hurricane frequency, but the long-term trend in a relationship is small, if it exists at all. He writes:

The marked increase in land-falling hurricanes in Florida and Japan in

2004 has raised questions about whether global warming is playing a role.... [The] observational hurricane record reveals large natural variability from El Niño and on multidecadal time scales, and that trends are therefore relatively small. However, sea surface temperatures are rising and atmospheric water vapor is increasing. These factors are potentially enhancing tropical convection, including thunderstorms, and the development of tropical storms. These changes are expected to increase hurricane intensity and rainfall, but the effect on hurricane numbers and tracks remains unclear.[24]

The reality that humidity ("atmospheric water vapor") is increasing could be explained by the fact that, periodically, weather conditions in the tropics produce an unusual number of days in which clear skies prevail, allowing more energy to be absorbed in the oceans. This may well explain why the ten-year period of time in which hurricanes increased in number in the Atlantic failed to be replicated in the Indian and Pacific Oceans. Rising ocean temperatures caused by sunlight penetration hence make it possible for tropical storms and hurricanes to form. However, there are times during which light energy from the sun is not able to penetrate the atmosphere and reach the ocean's depths. Normally, cloud cover is the most common obstacle for insolation (sunlight) reaching the surface of oceans and land, for that matter. Cloud cover, however, is not the only atmospheric property that has a high albedo (reflectivity) rating. Indeed, some forms of particulate matter (pollution and volcanic discharge) in the upper layers of the tropical atmosphere can impact global weather conditions by reflecting energy back into space. This energy never penetrates the oceans, so the earth's ability to disperse energy via the hydrologic cycle toward the poles is diminished. When this condition exists, global temperatures decline.

As a case in point, consider what happened to midlatitude weather conditions when Tambora, a volcano in Indonesia, erupted in 1815. It caused sufficient quantities of particulate matter to enter the atmosphere where energy was either absorbed or reflected back into space. The result was a significant drop in midlatitude temperatures over the next year, which historians and geographers call "the year without a summer."[25] During that year, there were periods of warmth, but they were invariably followed by cold spells in each of the summer months. Snow accumulated in parts of Massachusetts in June, and in Cabot, Vermont, on June 8, snow reportedly reached a depth of eighteen inches.[26] Southern states were not spared from the abnormal weather. Thomas Jefferson wrote a letter to Albert Gallatin from his home at Monticello in which he described local drought conditions as well as summer weather patterns that seemed more like a mild winter than a typical Virginia summer.[27] Particulate matter, primarily ash and sulfur aerosols, that spewed from Tambora did not act alone in causing the year without a summer, however. The year 1816 marked the midpoint of the Dalton Minimum, one of the sun's extended periods of low magnetic activity. It lasted from 1795 to the 1820s.[28] It is also thought that the earth wobbled, which affected the amount of direct insolation that entered the earth's atmosphere. The drought conditions in Virginia, especially combined with colder air temperatures, suggest that moisture from the Gulf was not abundant, which is symptomatic of a lack of hurricanes and tropical storms in the Gulf of Mexico. Indeed, Andrés Poey, the author of the generally accepted source for North Atlantic hurricanes between 1493 and 1855, identified six hurricanes in 1815 with one observed as far north as 40°.[29] Poey claims that only two hurricanes made landfall in the south Caribbean on the island of Martinique in 1816. However, in 2006, climatologist Michael Chenoweth conducted a thorough reexamination of Poey's conclusions and decided to reject one of the 1816 hurricanes and two of the 1815 hurricanes.[30] Nevertheless, no hurricanes made landfall in North America in 1816. There is no way of knowing about the frequency of lesser low-pressure centers that formed in the Gulf. However, given the dry conditions that Jefferson wrote about, one could assume that the humidity fed into Alabama, Mississippi, and Louisiana by low-pressure centers was greatly reduced in 1816. Had there been normal amounts of humidity drawn into the westerly winds that typically cross the Mississippi and Tennessee Valleys, coupled with the colder temperatures that made Virginia seem as though it was experiencing a mild winter, condensation would have accelerated and, consequently, there would have been more convective storms and less powerful frontal precipitation. These weather phenomena certainly would have caught former president Jefferson's eye. On the contrary, it was their absence that he wrote about to Gallatin. Deadly hurricanes do serve a valuable purpose in dispersing moisture and heat poleward.

When viewed from a satellite, hurricanes like Rita and Katrina are huge, white masses of circulating condensation. They, like volcanic ash and particulate air pollution, are natural reflectors of sunlight. It could be argued that the unusually high number of Atlantic storms of 2005 served as reflectors of the energy that would have been needed to produce the unusually large numbers of annual hurricanes that was predicted by the National Hurricane Center.[31] Consider that globally about 49 percent of the sun's energy reaches the earth's land and water surfaces. However, thick cloud cover associated with the rain-producing nimbus variety reduces the amount of energy that reaches the earth's surface to only about 10 percent. This means that Rita and Katrina not only dispersed energy poleward, cooling the Atlantic's tropical waters, but also their cloudy existence for days over the ocean and Gulf of Mexico either absorbed or deflected short-

Image 7.3. Sunlight reflecting off the surf. *Image courtesy of the author.*

wave radiation from the sun, thus depriving the Atlantic of energy and making it less likely that hurricanes would soon form over them. The overall impact of their existence actually served to reduce the amount of energy that was able to reach the ocean. Without insolation, the energy source for subsequent hurricanes was not available. This explains why there have been relatively few Atlantic hurricanes in the years since their devastating visits in 2005. However, the fact that clearer conditions have prevailed since 2005 suggests that sunlight is feeding energy once again into the tropical waters of the Atlantic. It is only a matter of time before destructive hurricanes once again form over the north equatorial current.

The rather complex relationship between atmospheric visibility, hurricane formation, and their relationship to climate change (absent human causation) in a modern context is a topic seldom discussed in the popular literature. Nevertheless and regardless of the cause of energy transmission into the seas, rising sea levels spell jeopardy for seaside settlements when conditions are right to develop storm surge and flooding. If there is a relationship between hurricanes and global warming, it is quite complex and deserves further investigation because the role that people play in causing hurricanes to form may be more paradoxical than current thinking suggests. Indeed, if humans pollute less and the atmosphere becomes "cleaner" or void of particulate matter, the north equatorial current may get even hotter. Hurricanes, which are produced by sunlight penetrating ocean depths, should more logically be theorized as contributors to global climate change because they are mechanisms of energy dispersal into zones of energy deficits (poleward places approximately above or below 40° north and 40° south, respectively).[32]

The processes for dispersing energy toward the poles through atmospheric and oceanic systems have played major roles in shaping the outer parameters of the human ecumene. Let us now focus our attention on how the Coriolis effect, which was mentioned earlier, is created by the earth's shape and rotation. The Coriolis effect helps shape hurricane alley while making potentially cold, undesirable places like Skara Brae in Scotland and Mount Sandel in Ireland attractive places to live. Under normal atmospheric and oceanic conditions, water flowing in the upper currents of the

Atlantic is deflected to the right and moves along the North American continent as the Gulf Stream and eventually deflects eastward across the North Atlantic where it is known as the North Atlantic Drift. Once the current, which is now heading south, has given off most of its energy relative to increasingly warmer landmasses to the east, less evaporation makes Iberia and North Africa semiarid to arid before the water body once again heats up under increasingly direct sunlight. By this time, the Coriolis effect has bent the current westward where, depending upon cloud cover and particulate matter in the atmosphere, it starts the cycle of energy absorption all over again as the north equatorial current flows west toward North America. As the circular movement of the Atlantic currents suggests, the Gulf Stream is an odd name for the current because it does not originate in the Gulf of Mexico. Nevertheless, it is a current characterized by comparatively warm water, and if it contains sufficient energy, it can fuel hurricanes. Although the Coriolis effect does shape the direction of ocean currents and, hence, the most likely places that hurricanes will make landfall, comparatively warm gulf waters and upper-level winds can cause them to enter the North American continent somewhere between Florida and Mexico. As late as September 1900, the Coriolis effect was assumed by meteorologists to be a powerful enough force to protect the western shores of the Gulf of Mexico. After all, there was proof of that assumption: low-lying *Nouvelle-Orléans* was founded by the French in 1718. New Orleans was still thriving after 182 years, so why get alarmed about intermittent storms associated with the tropical easterly wave? That assumption, however, was flawed with deadly consequences.

THE GALVESTON HURRICANE OF SEPTEMBER 8, 1900

The National Weather Service in the twenty-first century is empowered by technology to make predictions that it could not perform in its early years. The Weather Service began in 1870, when President Ulysses S. Grant created the agency under the auspices of the Army Signal Corps and the direction of the secretary of war. It was assigned the responsibility of making meteorological observations, especially on the seacoasts and the Great Lakes, and then relaying that information to other points on the continent and territories. The Weather Service's original name gives ample description of why it was founded: the Division of Telegrams and Reports for the Benefit of Commerce. Unfortunately, the US Weather Bureau, which was the name of the agency by 1900, was staffed by observers of varying degrees of relevant education and training. The equipment at their disposal was of modest capacity to detect immediate threats, so forecasts depended heavily on communication between various places. By knowing the typical direction of wind and weather systems, communication between observation points allowed personnel on the leeward side of the line of communication to know what kind of weather was moving in their direction. The use of barometers to measure changes in air pressure and anemometers to gauge wind speeds gave observers some ability to make good predictions. However, the bureau also made some terrible mistakes in failing to predict life-threatening storms. Such was the case in New York City on March 12, 1888. The weather for the city was predicted to be "colder, fresh to brisk westerly winds, [and] fair weather."[33] To the bureau's embarrassing chagrin, the next day delivered the Great Blizzard of 1888. Four feet of snow fell on Albany and twenty-one inches blanketed New York City and its boroughs. Some two hundred New Yorkers died, while elsewhere in the northeast, the storm claimed two hundred more lives.[34] Weather prediction was clearly not an exact science in 1900.

On the western shores of the Gulf of Mexico, few people feared weather systems from the east because the typical weather pattern was a west-to-east phenomenon. Besides that, the Coriolis effect would send any would-be storms to the Northeast. Weather observers in Cuba, the North American birthplace of modern hurricane forecasting, thought that the storm that had dropped several inches of precipitation on the island on September 5–6, 1900, was a threat to points west in the Gulf; but the Weather Bureau was convinced that the Coriolis effect would send the storm to the Northeast along the coast where the Gulf Stream would likely sustain it as a powerful rainstorm, if not a hurricane. However, at eleven thirty in the morning on

Friday, September 7, the Weather Bureau notified its observers in Louisiana and Texas that a tropical storm located south of Louisiana was slowly moving northwest. For Isaac Cline, the bureau's representative in Galveston, Texas, he would operate on the forecast spelled out in the telegram: "high northerly winds tonight and Saturday [the eighth] with probably heavy rain."[35]

The storm's trajectory, as given by the Weather Bureau, was on a track from Cuba. The seafloor extending away from the barrier island on which Galveston sits slopes downward to about five hundred feet below the surface and is no deeper than that for about two hundred miles toward Cuba. A common feature of coastal areas in which the sea floor slopes gently away is the barrier island. Composed of sandy deposits from weakened ocean waves before they reach the main shoreline, barrier islands are made up of calcium-rich sediments and are not very stable. Loose, sandy material is moved about by wind and wave action, forming dunes. Lagoons or inter-coastal water ways form between these islands and the mainland. The movement of tides in and out creates inlets that sometimes breach an entire island, so there is a mixture of fresh and salt water in lagoons.

Sitting north of Galveston Island is Galveston Bay. It is an estuary rich in plant and animal life. As an ecological community, Galveston Bay supports juvenile marine organisms such as shrimp, oysters, crabs, and copious fish varieties. The bay is a cornucopia of valuable natural resources. As a southern center of fishing and shipping, Galveston in 1900 was equally rich in economic activity and culture. In fact, forty-five steamship lines connected Galveston to the outside world, including the White Star Line that would later become famous as the owner of the "unsinkable" Titanic. The town was wealthy enough to have electric street cars, telephones, electricity in homes and businesses (some thirty-three years before the TVA was formed), three big concert halls, and twenty hotels, including the richly appointed Tremont. Its citizens were optimistically looking forward to the new century.[36] Given the town's encouraging economic future and the original name of the agency, it is no small wonder that the US Weather Bureau operated an observation station on the island. As previously mentioned, it was managed by Isaac Cline, an ambitious, smooth-talking "story teller" from Monroe County in east Tennessee. Cline had studied for the ministry at a small, two-year college located at nearby Madisonville. The self-trained meteorologist and his family, including his younger brother Joseph, made up part of the population of Galveston that the recently conducted census showed was just over twenty-nine thousand. Galveston's economic vitality was seen in its population growth; the city had grown by 30 percent since 1890.[37] However, despite being situated on a barrier island with its highest elevation at a few inches shy of eight feet, the city's material splendor invited residents to dismiss thoughts of potential harm from human-environmental interactions. On the morning of September 8, 1900, however, the bustling little city and it optimistic residents were in jeopardy from a force of nature that few of them could have imagined.

This wide expanse of relatively shallow water in the Gulf had been bombarded by sunlight for a number of days as a stationary high-pressure center settled in over the northern Gulf area, the Gulf States, the Midwest, and the Eastern Seaboard. The oppressive mass of stationary air stretched far out into the Atlantic. Clear skies permitted sunlight to superheat the Atlantic and the Gulf of Mexico.

Throughout the month of August, there were frequent convective rain showers over the island. Being situated at 29°16'52" north, the island is far enough south that there are scant few opposing and colliding air masses throughout most of the year. Most of the precipitation that falls on Galveston comes in the form of heat-generated rain events and not by frontal activity, although fronts are more common in the cooler winter months. However, the island is within the latitudinal range of the easterly wave that moves westward between 5° and 30° north at a speed of about 250 miles per day. The easterly wave is a weak low-pressure center (trough) that can take up to two days to pass over a place like Galveston.

The intensity of sunlight penetrating the Gulf fed humidity into the regional weather system. E. B. Garriot, the Weather Bureau's chief forecaster at the time, claimed that most of the eastern United States had suffered through the hottest August on record.[38] That was not much of a declaration because official weather records had only been kept since 1873. Nonetheless, the atmosphere was priming itself for even more energy transmission. Temperatures were in the one-hundred-degree range in a number of East Coast cities. Philadelphia reached 100.6°F at three o'clock in the afternoon on August 11. On that same day in New York City, thirty people died from the heat, and, sadly, three children who had sought relief from breezes on a fire escape fell to their deaths.[39]

In the Gulf of Mexico, the developing low-pressure center sent warning signs in front of its march toward Galveston. In the hours before daylight on September 8, Isaac Cline and his brother Joseph, who also worked for the US Weather Bureau, were awakened and troubled by the thudding sounds they heard through their open windows. Unusually loud waves were crashing onto the beach. This was peculiar because the normally placid Gulf produced little more than the gentle-wave action of a small lake.

Concerned about the crashing sounds, Isaac, as he reported later, rose from bed hours before daybreak and walked the few blocks to the beach. He observed large waves and a higher tide than normal. He also observed a strong north wind. Cline then went straight to his downtown office and checked the station's weather instruments. Despite his "feeling" that some-thing was terribly wrong, he decided that the evidence did not suggest a hur-ricane. In reality, the north wind should have told him that there was a powerful low-pressure center forming over the Gulf. Nature abhors a vacuum, and a hurricane or severe low-pressure center is indeed a huge vacuum. It was literally "sucking air" in from Texas as well as moist air from other parts of the overheated Gulf of Mexico.

As the morning wore on, hundreds of people came down to see the breakers. Children and adults enjoyed the cooler north wind while playing in the increasingly large waves. Intermittent blue sky and orange-hued clouds reigned overhead. One observer of the morning's activities at the beach was Walter W. Davis of Scranton, Pennsylvania. He was in Galveston on business, and as a resident of a landlocked state, he was not accustomed to seeing the ocean. When he heard people in his hotel talking about the huge beakers, he was compelled to pay a visit to the beach and see them crash ashore. As he recalled, "the site was grand at the time. I watched the waves wash out and break all of the shell houses, theaters, and lunch rooms, until I saw the waves were coming too close for comfort."[40] Even though the hurricane was sending forth plenty of warnings, Davis walked to his hotel through water choked streets, often knee high, ate lunch in the hotel's dining room, and then gave into his curiosity by leaving the inn and venturing to the bayside of the island. He stood on a high sidewalk and gazed in amazement as water from the bay invaded the streets before and behind him. He noted that he could see the water rising, and shortly he "became nervous" when the water careened over the sidewalk and washed over his feet.[41] He did not know it, but he and the population of Galveston were witnessing storm surge, and the eye wall of the hurricane was still dozens of miles out to sea.

Even at half past noon, there still were no hurricane warnings issued by the bureau. In his report to the Weather Bureau on September 23, Isaac Cline stated that the storm warning that had been issued from Washington, DC, was sufficient and that precautions to warn people had been taken. He claimed that the warnings he issued saved thousands of lives. However, the strength of his warning and his actions raise questions about the depth of his concern. As he wrote to the Weather Bureau:

> Storm warnings were timely and received a wide distribution not only in Galveston but [also] throughout the coast region. Warning messages were received from the Central Office at Washington on September 4, 5, 6, 7, and 8. The high tide on the morning of the 8th, with storm warning flying, made it necessary to keep one man constantly at the telephone giving out information. Hundreds of people who could not reach us by telephone came to the Weather Bureau office seeking advice. I went down on Strand Street and advised some wholesale commission merchants who had perishable goods on their floors to place them 3 feet above the floor.[42]

Those warnings were not nearly enough to save the city's residents from the unnamed hurricane's wrath.

Throughout most of the day and early evening, Cline or his staff recorded barometric pressure as well as wind speed and direction. Their record keeping in the midst of an emerging massive hurricane shows that they believed they were relatively safe. In his report, Cline recalled that he went to work at five o'clock in the morning and throughout the morning made several trips between his office and the beach. Torrential rain began falling at noon. He was concerned because upon his first visit to the beach that morning he noted that heavy swells, which were occurring at high tide, were overflowing the low places in the southern portion of the city. Despite his apprehension, his instruments did not tell him that a hurricane was developing. His reluctance to accept the fact that a hurricane was heading full force for his low-lying island was reinforced by other, less-reliable assumptions: "The usual signs which herald the approach of hurricanes were not present in this case. The brick-dust sky was not in evidence to the smallest degree."[43] Furthermore, he did not observe any nimbus clouds until after noon: "Broken stratus and strato-cumulus clouds predominated during the early forenoon of the 8th, with the blue sky visible here and there. Showery weather commenced at 8:45 a.m., but dense clouds and heavy rain were not in evidence until about noon, after which dense clouds with rain prevailed."[44] It was at about this time that Walter W. Davis had watched rising water from the bay.

While Isaac went home for lunch at three thirty in the afternoon, Joseph stayed at the office to send an update describing the flooding conditions to Washington. He discovered that the city's telegraph wires were down and that only one telephone line connecting the island to the outside world was working. Joseph was eventually able make a call to Houston and gave officials there this report: "Gulf rising, water covers streets of about half of city." That was the last report filed until after the storm subsided. Winds reached hurricane force at five o'clock. The highest wind speed recorded that day was 100 mph at six fifteen in the evening, but, according to Cline, the anemometer blew away at that time. Given the fact that the eye wall did not make landfall until after eight o'clock and that Cline described the wind after its passage to be of "greater fury than before," wind velocity at the storm's peak was probably in excess of 150 mph. Cline, however, estimated a maximum velocity of 120 mph. It is important to point out that Cline did use the term "eye wall" in his report. Nonetheless, it is possible to determine the time of the passage of the eye wall by noting Cline's description of atmospheric conditions as well as shifts in wind directions. He noted that after eight o'clock, there was a "distinct lull." He also wrote that the wind then shifted from the northeast and the northwest to the southeast and south after the lull.[45]

That Cline and anyone else for that matter survived a direct hit from a category 4 storm is astonishing. Having a well-built, multistory home located a few blocks from the beach at an elevation of 5.2 feet above sea level—just two feet short of the highest elevation on the thirty-mile-long island—Cline believed that he and his family were safe in it, and so did several dozen of his neighbors. When Cline went home for lunch at three thirty, he found that the water around his house was waist deep. By eight o'clock, forty to fifty people had taken refuge in his house. Just before debris (or a tornado) knocked the sturdy house from its foundation at eight thirty, Cline noted that water was fifteen feet deep at his residence, making a storm surge of just over twenty feet. Of the fifty or so people who went in the black torrent, only eighteen survived. Among those who perished was his wife.

When the house collapsed into the tide, Isaac was able to find his children and his brother. Together with several others, they climbed aboard a pile of floating debris, clung for their lives, and were pushed out to sea by waves and then washed back toward shore. Three hours, later they landed on someone's home. They were three hundred yards from where his house stood. By eight o'clock on Sunday morning, the wind was blowing from the south at twenty mph, but more importantly, the water had receded. Of the city's population of twenty-nine thousand people, the homes of twenty thousand of them, according to Cline's report, were destroyed. While Cline estimated the loss of lives to be six thousand, writer Eric Larson provides a higher estimate of at least eight thousand.[46] Because Cline had the responsibility of issuing warnings that would have saved lives, Larson contends that Cline had good reason to issue a conservative estimate of the loss of life.

SUMMARY

Isaac Cline concluded his report of September 23, 1900, with a comment that the high death toll was the result of waves. He further argued that more lives would have been saved had there been a sea wall built in the Gulf. In the wake of the storm, such a structure was constructed. However, as Cline pointed out in his report, airborne debris killed many people as they tried to flee the city—an impossible task because bridges were washed out and boats could not sail in the bay. At any rate, a sea wall would not have saved them from airborne projectiles. In the final analysis of the Galveston storm of September 8, 1900, it is doubtful that a sea wall would have protected the city. Walter Davis's description of bayside flooding shows that the island, which in effect was a sea wall for the bay, could not stop the water from rising.

Given that hurricanes are one of nature's means to disperse surplus energy into zones of energy deficit, they will return. The Gulf of Mexico is not immune from them. While developers can build homes and businesses out of wind-resistant materials and even place them on stilts, there are uncontrollable factors that make it nearly impossible to make any low-lying, coastal settlement in the Gulf completely safe. Cline's home was built on an elevated foundation, but it was debris or a tornado that knocked the house into the raging sea. Trees and other natural features can and will become high-velocity missiles or torpedoes.

Given our penchant for living near large water bodies and our confidence in technology to overcome nature's forces, human population will increase in places like New Orleans and Galveston.[47] At the risk of sounding like an alarmist, it is nevertheless important to state that, at some unknown time in the future, a storm will develop over the north equatorial current. It will be fed by energy from the sea through the hydrologic cycle. Atmospheric conditions will empower the storm to strengthen and move over the Gulf or the Atlantic shoreline following the Gulf Stream. It will be a category 5, and it will be a killer.

Chapter 8

LIFE AND DEATH IN AMERICA'S TORNADO ALLEYS

INTRODUCTION

Beginning just after 8:40 a.m. on April 27, 2011, multiple EF2, EF3, and even EF5 vortexes destroyed portions of towns and cities from south Alabama to Virginia.[1] At least 340 people were killed, including 238 in Alabama, and over one million homes and businesses were without electricity for days after the outbreak.[2] Those residents and property owners who suffered only power outages were fortunate because more than ten thousand homes and buildings were destroyed, and insurance claims were estimated to approach $5 billion, making this tornado outbreak the most expensive in US history.[3] Despite early predictions of tornadic activity as well as warnings, the rash of 288 funnels on April 27, 2011, became the third-deadliest outbreak in American history. A number of the first victims were asleep and did not know that the National Weather Service had issued warnings. Other victims had taken shelter in places that they thought were safe.

The 2011 spring tornado season took more lives in late May, and the register of the top-ten deadliest tornadoes changed. In the early evening hours of Sunday, May 22, residents of Joplin, Missouri, a city in the western Ozarks, were given a twenty-minute warning that a powerful tornado was headed their way. Blocking out the setting sun, the eastward-moving EF5 ripped a three-quarter-mile wide swath through the city of fifty thousand. Despite the early warning, the single vortex packing winds in excess of two hundred mph killed an estimated 142 residents while leaving another 750 injured.[4] By the end of May, it was understood that only one of the top-ten deadliest tornadoes in US history had exacted its toll in America's Great Plains, which observers assume is the only area of the country thought to be in Tornado Alley.

The 2011 storms were devastating, but, fortunately, their respective fatalities fell short of the number of lives lost in the outbreak of April 3–4, 1936, in which 454 residents of Mississippi and Georgia were killed.[5] The 1936 and 2011 outbreaks were exceptionally cruel to Alabama and other Southern and border states. The loss of life in each of those tragic events, however, was significantly lower than the number of lives taken by a rampaging storm that devastated Missouri, Illinois, and Indiana in March 1925. These numbers tell us that more people have been killed by tornadoes in the Mississippi Valley and the Deep South than in any other region in the world, let alone the United States. This somber fact, however, is more a product of population density than it is physical geography. If the urbanization and population density in the Great Plains matched that of the states situated along the Mississippi River, the loss of life and destruction to infrastructure in America's heartland would be disconcerting indeed.

Stretching from the flatlands of northern Texas through the rolling prairies of Oklahoma, Kansas, and Nebraska, the Great Plains states are potentially deadly places to live. Over one thousand tornadoes annually touchdown in the United States and most of them do so in the Plains states. However, because of the low population density in America's heartland, few people beyond the bands of storm chasers who study them even know about their existence. Nevertheless, cinematic images of Dorothy and her little dog, Toto, fruitlessly trying to flee the path of a Kansas tornado, and Bill and Jo (meteorologists in the 1996 blockbuster *Twister*) hunkering down under a wooden bridge (something a knowledgeable meteorologist would not likely do) have affected the popular imagination of Americans and have no doubt influenced their geographical knowledge of killer tornadoes.[6] These films do, however, offer a valid point: living in Kansas or anywhere else in Tornado Alley is potentially dangerous.

Despite forecasting and communication technology, tornadoes are still deadly, formidable opponents. In addition to the loss of life, billions of dollars in storm-related damage has affected the United States. The Oklahoma tornado of May 3, 1999, caused $963 million in damages (constant 1997 USD). That storm's monetary toll on the United States is not quite half of the relative cost of the country's second-most-expensive tornado. On May 27, 1896, a vortex tore through St. Louis, Missouri, and East St. Louis, Illinois, leaving death and destruction in its wake. The dual-state twister killed 255 people.[7] It is estimated to have exceeded $2.2 billion when adjusted by inflation and economic expansion.[8] As with hurricanes, these storm-related costs ripple through the country's economy as insurance companies increase premiums to offset their losses. Even government interventions require additional debt and taxes to pay for damages. As more people settle in dangerous places, the number of lives lost and costs to rebuild infrastructure will likely increase. While some housing structures are better than others, few building designs and materials are able to withstand a direct hit from an F4 or F5. Even if a building can resist winds approaching three hundred mph, there is little it can do to absorb the impact of airborne automobiles, bricks, and other debris that add to a tornado's destructive and killing powers.

There is much to know and appreciate about tornadoes. The manner in which humans cope with them is actually a relatively new activity. Tornadoes seldom occur in Africa, Asia, or Europe, so our ancestors did not leave us a great deal of accumulated knowledge on how to cope with them.

PREDICTING TORNADOES

Americans, true to the human ecology paradigm, have adopted technologies to issue warnings of impending, life-threatening weather events. Rudimentary attempts to understand and predict tornadoes were made in 1882 by Lieutenant John Park Finley of the Army Signal Corps. By 1884, he had established fifteen rules for tornado forecasting. However, Finley's tornado research took a bad turn when, in 1887, General Adolphus Greely became chief of the corps. The new chief decided that it was impossible to accurately predict the exact location of touchdowns, so he ordered the use of the word "tornado" stricken from public forecasts, even though issuing them

could have been based on a probability scale or a statement such as "conditions are conducive to tornado formation."[9] After Greely's decision to
remove tornado forecasting in 1887, little was accomplished in imple-
menting tornado warning systems until Franklin Delano Roosevelt's last
term in office. In 1943, the Weather Bureau implemented experimental tornado warning systems for a limited number of locations: Wichita, Kansas;
Kansas City, Missouri; and St. Louis, Missouri. In June 1944, weather
observers in those cities began also to include thunderstorms, hailstorms,
lightning, and high winds in their warnings.[10] It was not until March 25,
1948, that the military and the US Weather Service were convinced that it
was possible to forecast a tornado for a general location. Using the basic
empirical approach devised by Finley, Major Ernest J. Fawbush and Captain
Robert C. Miller of the US Air Force predicted a touchdown for Tinker
Air Force Base near Oklahoma City. Their feat was even more impressive
because just five days earlier, on March 20, a tornado had destroyed some
thirty-two military aircraft causing $10 million in damages on the same
base. The idea that a specific location like Tinker could suffer two tornado
touchdowns in less than one week was not thought possible. Nevertheless,
the Fawbush and Miller forecast had enabled base commanders to implement a tornado safety plan, and, although the tornado of March 25 caused
$6 million in damages, no lives were lost.[11]

Despite their success in saving lives, it took five years and hundreds of
more lives to force Congress to create a national warning system. Tornado
outbreaks in Arkansas, Missouri, and Tennessee killed over 150 people on
March 21, 1952. As a result of an ensuing public outcry for a national tornado warning system, Congress authorized the creation of the Severe Local
Storms (SELS) Center within the Weather Bureau. It became operational
in June 1953, but it was not soon enough to save hundreds of lives lost that
same year.[12] On May 11, 1953, a tornado killed 114 people in Waco, Texas.
A few weeks later, in early June, a cold front moved across Michigan's Lower
Peninsula. With the jet stream creating an upper-level wind shear, a tornado
formed and touched down at Flint, Michigan, on June 8 killing 115 residents. The same system then travelled eastward across Lake Ontario and
Lake Erie where on June 9 it produced another powerful vortex that tore

through Worcester, Massachusetts. This time the death toll was lighter; only ninety were killed.[13]

Thanks in large measure to the implementation of technology to predict and then warn residents of a potential threat, lives have been saved. Until the outbreaks of 2011, only two of the twenty-five deadliest tornadoes occurred since the national warning system was established in 1953.[14] Since that year, most of the research on tornadoes has been focused on developing technology to make accurate predictions and warn people in sufficient time for them to take shelter. Nevertheless, there is room for improvement in warning systems and building construction, although few existing materials and structures can resist the projectiles hurled at them by an F4 or F5 tornado (the F-Scale is described in the next section of this chapter). Indeed, even concrete sidewalks and asphalt roads have been uprooted and sent flying across the landscape. All the same, Doppler radar and early warning systems have saved perhaps thousands of lives from powerful tornadoes.

Consider the case of the F5 tornado of May 3, 1999. There were more than one dozen tornadoes that touched down on that spring day in Oklahoma. This particular funnel stayed on the ground for hours, strengthening as it travelled northeast toward Oklahoma City. By the time the vortex reached Midwest City, a suburb of the state capital, it was massive. The F5 produced a superlative 318 mph hour wind velocity, the highest wind speed ever recorded. Continuing to move in an east-northeast trajectory, the colossal tornado mercifully ripped its path east of the state capital. Once the swirling funnel passed Oklahoma City, it skipped over parcels of farm and ranch lands. The killer of forty-four Oklahomans followed the Will Rogers Parkway (Interstate 44) until it dissipated before reaching Tulsa, but before dying, it leveled a strip mall at Stroud.

While the death toll from this meteorological event is small compared to the 695 lives lost in the Tri-State Tornado of 1925, the fatalities could have been even lower. Failure to get the word out was not the case in Alabama and six other southern states in April 2011, nor was it a problem in Oklahoma on May 3, 1999. Not only were Oklahomans warned through official channels, including municipal sirens, television news teams followed and filmed the monster as it approached and then leveled population centers. At the time, I was living less than one mile from Interstate 44 in Ottawa County, Oklahoma. A friend and colleague of mine lived in Claremore, a city just east of Tulsa. With our eyes glued to our respective televisions, we spent most of the evening on the phone discussing the tornado as we watched it make its way toward us. While the vortex did not reach either of us, we nevertheless experienced high-velocity winds, lightning, and torrential rain. More importantly, we had ample warning to find shelter, which, because of the severity of damage caused by the vortex, would have had to have been in an underground storm shelter. Moving to a central room in either of our houses would not have saved us. Even ranch homes made out of sturdy brick were leveled. Building technology certainly has its limits in withstanding a storm of that magnitude, but fortunately most people had ample time to find shelter or get out of the storm's path. Sadly, however, many of the forty-four Oklahomans who lost their lives that day believed that they were safe, for they had taken refuge in what they presumed to be storm-safe shelters.[15]

Some of the victims who were killed or injured had taken shelter in one of the most dangerous types of places one can find in the path of a tornado. Making a decision to hide under a highway overpass is perhaps understandable. Most of us who have travelled interstates have witnessed vehicles parked under bridges during intense storms, especially those that produce hail. Motorcyclists too are attracted to overpasses to escape torrential downpours. The fact is the restricted space created by the bridge and the ground actually channels higher velocity winds and debris. It is the same effect as placing one's thumb over the end of a running water hose. Because the thumb restricts the space at the terminal end of the hose, the velocity there increases so the volume of water traveling through can remain constant. The same principle applies to air and winds. The safety of an underpass is hence an illusion, and it is one that should be avoided in the presence of a tornado. While there is no guarantee of safety from any tornado that passes over one's outdoor location, the chances of survival are much greater in a low-lying ditch. However, many ditches in the southeastern United States are often wet and serve as the homes of cottonmouths or water moccasins. Clearly, it is prudent to watch where one takes refuge.

Oklahoma and other central plains states may form America's most famous tornado alley, but their relatively low population densities have been the primary reason why their death toll has been smaller than the region with the greatest loss of life. To more fully appreciate that statement, it is helpful to take a longer look at the taxonomy and geography of tornadoes. This information sheds light on why the South and the states along the Mississippi River, including its major eastern tributaries, form North America's deadliest portion of Tornado Alley. To illustrate that point, this chapter includes data on deaths resulting from historic storms. More in-depth historical geographies are also presented on the Tri-State Tornado of 1925 and the Mossy Grove tornado of 2002. In North America's central and eastern midlatitudes, there are few places completely safe from major tornadoes. As we will see in the next section, all tornadoes are not necessarily killers. There are precautions one can take to ride out some, while others require underground shelters or evacuation from the path of the vortex. However, since the trajectory of a tornado is liable to shift directions by a few degrees, evacuation may mean traveling several miles.

CLASSIFYING TORNADOES

As we saw in earlier chapters, tornadoes formed by frontal activity (the non-hurricane variety) are generally located in places far removed from the natural utopias marked by our ancestors thousands of years ago. This is not to say that Asians, Native Americans, Europeans, and South Africans did not know of their existence. Some of them certainly encountered vortexes and survived, but knowledge of tornadoes was somewhat localized. Also, the ability of people to cope with these meteorological events was limited and most likely assigned to the realm of geotheokolasis, which Thomas P. Grazulis succinctly called the "direct wrath of God."[16] In the case of Native Americans, few of them lived in the Great Plains until they acquired horses left behind or set free by Spaniards in the sixteenth century. In other words, our population growth and settlement into the "damage path" of tornadoes is, in the scope of European and Asian histories, a fairly recent situation.

Nevertheless it is possible to find descriptions of tornado activity in ancient writings in the Near East. In the Bible, Job 37:9 describes north and south winds forming whirlwinds capable of scattering things. Isaiah 42:15–16 describes how God will use a "whirlwind" to scatter Israel's obstacles. Depending on whether or not one was an Israeli, Isaiah's description can be seen as an example of either geotheokolasis or geotheomisthosis. Following the biblical lexicon, John Winthrop, the Puritan governor of Colonial Massachusetts, used the term "whirlwind" to describe the damage caused by a 1643 storm, which was most likely an F1 or F2 tornado. The vortex killed one Native American, blew down many trees, and lifted up a meeting house in Newbury; he did not use the term "tornado" to identify the calamitous event.[17] In 1680, Reverend Increase Mather also used "whirlwind" to describe a similar event in his book *Remarkable Providences*.[18] The use of "tornado" to describe a violent whirlwind was not firmly in place in the American lexicon until the middle of the eighteenth century when Reverend Joseph Emerson claimed that a "terrible tornado with shocking thunder" touched down at Groton, Massachusetts, in July 1748.[19] Arguably the next great leap forward in the taxonomy of tornadoes did not occur until 1971 when Tetsuya "Ted" Fujita (1920–1998) developed his famous classification system.

Fujita, who was a professor at the University of Chicago, teamed with Allen Pearson of the National Severe Storm Forecast Center (NSSFC) to develop the Fujita or F-Scale. They based their taxonomy on measured wind speed and the extent of damage caused by storm events.[20] The fact that such a scale did not exist until 1971 is testimony to the fact that population densities in areas prone to suffer tornado touchdowns were, by then, sufficiently large enough to cause widespread, public concern about them. As a means of understanding tornadoes, the Fujita Scale had its limitations, so in 2007, a national forum of atmospheric scientists developed an Enhanced Fujita Scale (EF). The scientists felt it was necessary to consider the quality of construction material in assessing wind damage (see table 8.1).[21]

Armed with Fujita's useful scale and the rationale for its development, let us now shift our attention to a more in-depth discussion of the geography of tornadoes.

Table 8.1. Original and Enhanced Fujita Scale

F0: Gale tornado. This mild tornado produces winds of 40–72 mph, damage to some chimneys, and some broken tree branches. EF wind speeds: 65–85 mph.
F1: Moderate tornado. This slightly more destructive vortex produces winds of 73–112 mph; it produces damage to shingled roofs, mobile homes are knocked off foundations, and there is considerable damage to detached garages. EF wind speeds: 86–110 mph.
F2: Significant tornado. While the winds (113–157 mph) of this level of tempest are destructive to nonreinforced structures, it is possible to ride out the storm in the interior of most well-built houses. In such a storm, roofs are torn off frame houses (wooden walls and rafters), mobile homes are destroyed, and large trees can be uprooted. Light objects are transformed into airborne projectiles. EF wind speeds: 110–135 mph.
F3: Severe tornado. This is potentially a killer tornado. Winds range from 158 to 206 mph; roofs are torn off all houses, exposing even the interior rooms, which are thought to be among the safest places in a home; trains are overturned, and most trees in the storm's path are uprooted. EF wind speeds: 136–165 mph.
F4: Devastating tornado. This killer produces winds between 207 and 260 mph. Only basements of houses are safe because even well-constructed homes are destroyed. Household appliances are thrown long distances. EF wind speeds: 166–200 mph.
F5: Incredible tornado. An F5 is the strongest tornado ever witnessed. It produces winds of 261–318 mph; strong houses are lifted off foundations and disintegrated. Cars are thrown more than 100 meters, and even steel-reinforced concrete buildings are badly damaged. EF wind speeds: > 200 mph.

While both the F-Scale and the EF-Scale are accurate assessments of tornadoes, the EF-Scale factors in the degree of storm damage. By aligning wind speeds to reflect the amount of damage produced by the storm, we can better understand the tornado.

GEOGRAPHY OF TORNADOES

While tornadoes infrequently develop in Europe, South Africa, and Australia, they are actually quite concentrated in a fairly small amount of space. Most of them occur in the American Great Plains, the southeastern United States, and the states along the Mississippi River and its major tributaries. As with hurricanes, certain meteorological conditions must be present in order for them to form. To develop, a tornado must have rapidly rising air associated with an intense low-pressure center formed along the trough of an aggressive cold front. As air rises, it does so in a counterclockwise rotation in the Northern Hemisphere. Subsiding air or high pressure sinks in a clockwise rotation. When a low-pressure center has enough instability exacerbated by an upper-level wind shear, which creates an even stronger vacuum effect, a tight spiral or rotation can form. It is in this situation that a tornado is most often born. Still, the probability of a storm producing a tornado is low. One in one thousand thunder storms features rotating winds and sufficient size to be classified as a supercell. Just one in five or six supercells spawn tornadoes.[22]

Low-pressure centers form across much of the earth's landmasses, but comparatively few of them produce these monsters. Conditions that produce tornadic activity are best met in the Western Hemisphere and specifically in the northern midlatitudes (30° to 60° north). In this belt of latitudes, cold, dry air masses in the spring, early summer, and fall often spread out from Canadian high-pressure centers and enter the United States over the Great Plains. This situation is especially dangerous when warm, moist air also enters the country from the Gulf of Mexico. The jet stream forms out along the edges of the cold fronts. Its high velocity winds have the capacity to create an upper-level wind shear that increases instability in low pressures that form below them. An aggressive cold front is actually a sinking mass of dense, dry, comparatively cold air. As such, ground friction causes the lower portion of the cold air mass to move slower than the upper part of the front. The effect of surface friction on a mass of cold, rapidly advancing air is similar to using only the front brakes on a bicycle. The rider will pitch forward off the bike. In the case of a cold air mass, the upper portion pitches forward, creating a squall line of heavy, foreboding clouds in advance of the actual cold air slowly dragging along the ground. This situation can also cause a wind shear and spawn a tornado.

While tornadoes may form in any month of the year, they are more likely to form in the spring and early summer months when opposing air masses are more common in the midlatitudes. Fall, too, is a season of weather transition, especially in the more humid southeastern portion of the country. Even the rugged uplands of southern Appalachia are not immune from autumn tornadoes. In fact, a vortex (this one generally regarded by the National Weather Service as a powerful F3) touched down on November 10, 2002, in the small Morgan County, Tennessee, community of Mossy Grove. The tornado left seven people dead. A more detailed discussion on the Mossy Grove tornado will be presented later in the chapter.

For the time being, it is worth exploring why other midlatitude locations are not as vulnerable to tornadoes. In western Europe, for example, the North and Baltic Seas, which lie to the north of the continent, are not ideal for creating cold, dry air masses. On the contrary, they contribute atmospheric humidity to the northwestern portion of the continent. To the south lies the relatively cool Mediterranean Sea. It contributes little atmospheric moisture to Europe, and much of it precipitates out on the slopes of the Alps. In Asia, the Himalayas and Tibetan Plateau also serve as orographic barriers to invading moist air from the Indian Ocean. As air rises up the southern slopes, condensation occurs, and by the time the air mass meets high pressure from Siberia, they are similarly dry.

The Southern Hemisphere, known by geographers as the water hemisphere, actually has few land areas in the southern midlatitudes. Only southern Australia, New Zealand, South Africa, Chile, and Argentina lie in that belt of latitudes. These places, like the Great Plains of the United States, have comparatively low population densities, so the few tornadoes that touchdown in these countries often go unnoticed by societies not affected by their power.

The overarching issue created by a combination of population growth, expanding settlements, and filling in spaces in the ancient human ecumene will place future generations in harm's way. As is shown below, most of the deadliest tornadoes have occurred in places that have greater population densities than the Great Plains, America's most famous portion of Tornado Alley. However, these places, like the southern Great Plains, are experiencing population growth. As more people move into these vulnerable places, the death toll will inevitably rise, even with improved warning systems. The only solution will be to construct enough underground storm shelters to house hundreds of thousands of people.

This point is perhaps best appreciated by the geography of deaths resulting from tornadic activity. According to the National Geographic

Society, a major sponsor of a good many studies on tornadoes, Texas leads the United States in the average number of annual tornadoes with 120.[23] The greatest loss of life in Texas from a single outbreak of tornadoes occurred on May 11, 1953, when they killed 114.[24] This number is high in comparison to the average number of deaths by tornadoes per year in the United States, which sits at eighty. However, 114 deaths is small compared to the loss of lives in Alabama on March 21, 1932, when some 332 people died from a single outbreak. As we discussed at the beginning of the chapter, the death toll from the outbreak of April 27, 2011, is also much higher than the annual average number of Americans killed by tornadoes. However, even that large number is diminished when compared to the loss of life that occurred in Tupelo, Mississippi, and Gainesville, Georgia, on April 5–6, 1936. In that outbreak, 454 people were killed.[25] On April 3–4, 1974, a rapidly moving cold front left behind a swath of death and destruction that stretched from Alabama to Xenia, Ohio. Three hundred fifteen people were killed in that outbreak.[26]

With exception of Xenia and a few other places, the majority of those killer tornadoes touched down in mostly small settlements and across farm and ranch lands. The number of deaths could easily have been much higher if they had occurred in Atlanta, Birmingham, or Jackson. As population swells and more cities appear on these rural landscapes, it is conceivable that the Tri-State Tornado of 1925 could lose its infamous status as America's most deadly tornado.

The concepts of geotheokolasis (God using nature to punish those who sin) and geokolasis (nature punishing those who stray from living in harmony with nature) have provided human societies with psychological frameworks for coping with the forces of nature. Consider the following: one hundred years ago, the average person would have likely believed that an F5 tornado that killed nearly seven hundred people was an act of God. In contrast, in the twenty-first century, such an event has become a hot political issue because a large number of even secular people now believe that nature punishes societies for polluting the atmosphere, lands, and seas, and killer tornadoes fit that belief. This was certainly the case in the minds of many observers in regard to a perceived relationship between the

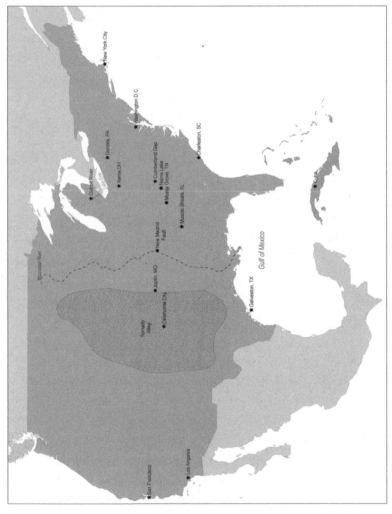

Map 8.1. Tornado Alley. *Image courtesy of Geraldine Allen and Clayton Andrew Long.*

destructive power of hurricanes like Andrew and Katrina and the notion of human-caused global warming (see chapters 4, 7, and 10 for discussions on those topics). It is not hard to imagine ancient peoples offering sacrifices to vengeful gods to remove the threat of killer storms, and it is similarly to be expected that many well-educated people today will turn to government to enact some legislation to stop people from doing the things believed to cause nature to behave in destructive ways.

Even among those who study weather and climate, there is disagreement about the human factor in meteorological events. A geographer named Wallace Akin who authored *The Forgotten Storm: The Great Tri-State Tornado of 1925* certainly holds a geokolasis view of the relationship between human actions and the frequency and intensity of hurricanes and tornadoes. Notice, in the following excerpt, how he uses "gluttony," one of the seven deadly sins, in his assessment of the current state of human-environmental interactions:

> I, as do most scientists, believe that global warming is increasing the frequency of severe storms, and this is a situation little understood by the public and their elected officials, who tend to focus only on temperature changes. But it is more than just warming. Our gluttony for energy is loading the atmosphere with pollutants and increasing the greenhouse effect. This subject is far too complicated to discuss briefly, but be assured that we can expect more frequent and more violent acts of nature, more hurricanes, more heavy deluges of rain with resultant floods and mudslides—and more frequent super tornadoes.[27]

In the wake of the killer outbreak of April 2011, Brian Williams, anchor of NBC's *Nightly News*, asked Jim Cantore, a reporter from the Weather Channel, for an explanation of the destruction. He asked Cantore what we had done to cause these events, which Williams insisted were historically unusual.[28] As was discussed previously in chapter 7, Max Mayfield, director of the National Hurricane Center, meteorologist William Gray, Jeffrey Kluger, and Kenneth Trenberth dispute the issue of human causation in storm intensity and frequency. Indeed, an increase in sunny days in the tropics resulting from lower levels of particulate matter (a form of pollution) and cloud cover could increase ocean surface temperatures that accelerate evaporation and, hence, humidity, one of the key factors in tornadic activity. As discussed earlier, tornado formation is a function of humidity, an aggressive mass of cold air, and an upper-level wind shear, which can be provided by the jet stream. That high-velocity wind resides in front of a mass of cold air. It is formed by atmospheric instability that causes relatively warm air to rapidly rise. When mapped, the jet stream and cold fronts form Rossby waves, which look like a cracking whip with ridges and troughs. Low-pressure centers and supercells form in troughs. The fact that deadly tornadoes struck as far south as Tuscaloosa, Alabama, indicates that an invading mass of cold air from the north was the culprit, not human-caused global warming. At any rate, Wallace Akin's assessment of human culpability in causing dangerous tornadoes is not as widely embraced among scientists as he would have us believe. While the validity of associated theological beliefs is beyond the scope of the present discussion, it is nevertheless interesting how some people, including learned scholars such as Akin, are quick to blame "sinful" humans for calamitous events caused by nature. In that regard, little has changed in human history.

At a minimum, however, humans are culpable for making decisions to live in harm's way. Just as many children have heard the refrain, "If you play with fire, then you are going to get burned," that same logic can be applied to living in the shadows of an active volcano. At some time in the future, residents, or some of their descendants, will be killed when it erupts. Living in Tornado Alley is also precarious, although warning technology has greatly increased the likelihood of surviving a tornado. In the 1950s when the tornado warning system was put in place, the average warning time was five minutes, but today the warning time has more than doubled to between thirteen and fifteen minutes.[29] Despite this improvement, killer tornadoes often deviate from places where one would expect them to touch down (i.e., an open prairie). Also, households do not always have a radio or television through which to receive warnings. There are still many municipalities that do not have sirens, and if they do, residents may not know what to do if they hear one breaking through the noise of the midafternoon or the silence of the night. By the time they tune in to a radio broadcast for details, it may be too late to take shelter. Because of distances from municipalities that may have sirens, rural residents are at an even greater disadvantage in receiving tornado alerts. An approaching tornado may knock out electricity before rural homes in its path are able to receive warnings through television or radio. A battery-powered radio would be most helpful in such a situation, but if the rural area has only a local AM station that shuts down at dark, these rural folk are in a precarious situation indeed, unless the tornado conveniently arrives during daylight hours.

THE TRI-STATE TORNADO OF MARCH 1925

Such was the case at the noon hour on March 18, 1925. It was an unusually warm day for that time of year in the Ozarks of southeastern Missouri. The pleasant midday temperature hovered around 65°F. Few locals would have thought that the deadliest tornado in American history would start its destructive life in these forested Ozark uplands. The small hills and ridges rise some four hundred feet above valley floors, where intermittent springs gush out water from limestone caverns. The rugged landscape in that part of the Show Me State certainly does not resemble the grassy plains of Dorothy's Kansas. Its uneven topography bears a striking resemblance to its sister uplands of south-central Kentucky. In fact, the culture of the Ozarks and its land-use practices were introduced by immigrants from Appalachia over the last two centuries. The earliest settlers were likely familiar with whirlwinds or tornadoes. Many residents had constructed storm cellars that doubled as fruit cellars for storing canned goods and other perishables. After March 18, 1925, home owners without a storm shelter wasted little time constructing underground sanctuaries.

The killer vortex began its 219-mile trek of destruction just before one o'clock in the afternoon in such a place near Ellington, Missouri, a small town in Reynolds County. The first person to recall seeing it was a rural mail carrier as he travelled his route just outside of Ellington.[30] The tornado, though certainly massive, moved along the earth's surface at an impressive speed of sixty to seventy mph. The average speed of a North American vortex is a mere thirty mph, about half the speed of the Tri-State Tornado.[31] Observers noted that at times the typical shape of a tornado was not visible. As the rolling mass of cloud and debris appeared over the hills to the west of the small Ozark village of Annapolis, few survivors recalled hearing anything more than a thunder clap before the sky went dark in a matter of minutes. Despite the damage leveled on the physical infrastructure of Annapolis, which was nearly wiped out, the storm left behind only four dead.

One of the Annapolis survivors was a traveling business man named C. E. Pyrtle. After having a leisurely lunch, he got into his car around 1:15 p.m. to leave the small town. As he started the engine, he became alarmed when he noticed a huge, black cloud appear in the sky above the street in front of him. He abandoned the car for a presumably safer place, but before he could get very far, he was knocked unconscious. When he awoke, he had a throbbing headache and a sense of bewilderment, for most of the town that had hosted him for lunch was strewn about in piles of rubble.

Since it was still in the middle of March, many of the homes and businesses were warmed by smoky coal- or wood-burning stoves. Even on that

Image 8.1. One of Joplin's decimated neighborhoods. *Photo courtesy of Lane Simmons.*

Image 8.2. Joplin High School. *Photo courtesy of Lane Simmons.*

Image 8.3. A Joplin home. *Photo courtesy of Lane Simmons.*

Image 8.4. A Joplin stone building. *Photo courtesy of Lane Simmons.*

Image 8.5. Tri-State Tornado touched down near this field. *Photo courtesy of Nick Cockrum.*

Image 8.6. Ellington, Missouri, as it appears today. *Photo courtesy of Nick Cockrum.*

balmy day, many of the residents had hot fires in their stoves to chase away the morning chill or to prepare meals. Because it was the end of the lunch hour, many of the stoves were still red hot. When the tornado ripped through homes and businesses, fiery coals were unleashed to consume wooden timbers that were lying in piles. Sadly, many of the injured who had survived the storm's initial onslaught were burned alive. Pyrtle described how he and other survivors rescued a sickly older woman and her caretakers before a fire took her home.[32] The same fate befell others as the storm continued its deadly charge. While the loss of even one human life to an act of nature is a tragedy, the Annapolis death toll could have been much worse. Other cities that day were not as fortunate as Annapolis.

When the rolling mass of clouds descended the bluffs that separate the Ozarks from the Mississippi River floodplain, some witnesses described the storm that carried pulverized debris as twin vortexes. Witnesses also commented on the various shades that appeared in the boiling clouds. Yellow, gold, black, and blue were among the many hues that appeared in the mass of rapidly moving condensation. Being nearly spring, the colors were created by airborne debris made up of various flowers, grasses, leaves, soil, and timbers from destroyed homes and barns, all having been sucked up into the vortex. By the time the storm crossed the Mississippi River north of Cape Girardeau and entered Illinois, thirteen Missourians were dead, sixty-three victims were injured, and an estimated $564,000 (1925 USD) worth of property was destroyed.[33]

Southern Illinoisans share more cultural characteristics in common with Southern folk than they do their Northern neighbors who live in the Chicago land area. The southern portion of the state is also warmer and more humid than the Northern reaches of the state. Indeed, Southern cities like Paducah, Kentucky, are dozens of miles north of the small farm settlements that dot the gentle landscape of southern Illinois. The first town in Illinois to feel the wrath of the historic 1925 vortex was Gorham. At the time, the small city had a population of nearly five hundred people. The tornado had already left an eighty-five mile swath of destruction in Missouri, but it seemed to strengthen in southern Illinois. In the lowlands near Gorham, it pulled up grass by the roots. As was the case in Annapolis, the vortex tore through Gorham with virtually no warning. The weather throughout the morning was rainy, but at 2:35 p.m., the tornado struck. The moment of its arrival was well documented; electric clocks stopped running at that minute. Among the thirty-seven people killed in Gorham were a number of children who died as their school collapsed on top of them.[34]

The next town that lay in the tornado's path was Murphysboro. Whereas Annapolis and Gorham were small settlements that served surrounding farm families with a modest assortment of goods and services, Murphysboro was a bustling little city with a variety of businesses and manufacturers, including Brown Shoe Company and a leading processor of silica named Isco-Bautz. Not only did the tornado exact a heavy toll in lives lost, it devastated the economy for decades to come. In total, one hundred blocks were leveled by the vortex, and, as was the case in Annapolis, fire consumed many areas that were left partially standing. The death toll in Murphysboro was staggering. Some 234 people were killed and 623 were injured.[35] Wallace Akin estimated that 40 percent of the city was destroyed.[36] Akin also estimated that the tornado destroyed 30 percent of

Desoto, a small mining town located northeast of Murphysboro. These percentages pale in comparison to Parrish, Illinois. Ninety percent of it was destroyed. Even greater still was the physical toll on Griffin, Indiana—it was completely smashed.[37] None of the eighty to ninety structures in Griffin were undamaged. Some 60 percent of the residents were either killed or injured. Seven of those who survived the onslaught of the vortex were subsequently burned to death. They had entered the Kokomoor restaurant for late afternoon snacks and coffee. When the tornado knocked the eating place to the ground, the seven were heard trying to dig out from the debris that covered them. As people worked to free them from the wreckage that once was the Kokomoor, fire from the restaurant's stoves spilled out onto the timbers that lay all around them. Flames caught on the rubble and little could be done to save the trapped and screaming victims.

To make matters worse for those who survived the initial damage, the storm caused torrential rain to fall continuously for the next twenty-four hours. The Wabash River that drains the valley in which Griffin sat, along with the local water table, rose higher and higher as the hours passed. The valley quickly became saturated, and the debris field was inundated with slow-moving flood water. Griffin's streets were submerged beneath a foot of water. The governor had paid a visit to the damage scene but was nearly stranded by the rising water. Griffin, Indiana, was a victim of winds, fire, and water.[38]

The Tri-State Tornado of March 18, 1925, was America's most deadly vortex. When it was over, the death toll numbered 695 with over one thousand people injured. Its track was the longest on record, but perhaps the most important aspect of the tornado that we should ponder is the fact that it was born in an unlikely place: the Ozark uplands of southeastern Missouri. The vortex finally dissipated at around four thirty in the afternoon just southwest of Petersburg, Indiana. It was on the ground for three and a half hours. Given the fact that the tri-state area is much more populated today, one can only imagine how much damage a modern version of such a storm would cause. Perhaps even more importantly, the Tri-State Tornado should tell us that one does not have to live in Dorothy's Kansas to experience the full force of nature's most intense storm. In the next section, we will explore this point further by describing destructive tornadoes that rip through open country, towns, and cities located east of the Mississippi River.

The Mossy Grove tornado of November 2002 was a killer F3, and it struck at night in a seemingly well-protected valley in the Appalachian Mountains of east Tennessee.

MOSSY GROVE AND THE OUTBREAK OF NOVEMBER 11, 2002

Situated five miles from the small town of Wartburg, Mossy Grove is a bedroom community nestled in a valley on the eastern escarpment of the Cumberland Plateau in upper east Tennessee. Most of the residents who live in Mossy Grove work for the state penitentiary at Petros or Morgan County Schools, or they commute twenty miles into Oak Ridge. Being situated in a valley, Mossy Grove is in a seemingly safe place.

The Mossy Grove tornado was a lone F3 vortex that plummeted from the sky above the plateau. Indeed, the tornado of November 11, 2002, that killed seven people in that Morgan County village was just one of seventy tornadoes that played havoc with the lives of people, plants, and animals on that midautumn Sunday. While fall tornadoes are less common than their spring counterparts, they are nevertheless products of major frontal activity that often produce multiple vortexes. The seventy tornadoes that touched down on November 11 left paths of destruction in seven states, and all of them were east of the Mississippi River.

This outbreak was the largest single outbreak of tornadoes since November 1992 when ninety-four tornadoes touched down in thirteen states.[39] Again, it is important to stress the fact that autumn tornadoes are more likely to form in the humid southeastern portion of the country while the Great Plains and the Midwestern states are more likely to be hit by tornadoes in the spring and early summer months.

Mossy Grove and the outbreaks of November 11, 2002, deserve attention for several reasons. Being situated on the eastern escarpment of the Cumberland Plateau, Mossy Grove's location does not suggest that the vil-

Table 8.2. Deaths and injuries Resulting from November 11, 2002, Outbreaks[40]

State	Deaths	Injuries
Tennessee	16	55
Alabama	12	50
Ohio	5	21
Pennsylvania	1	19
Mississippi	1	55
Georgia	0	7
Indiana	0	3

lage is in harm's way from a tornado. The ridges, which are actually the tops of the locally dissected plateau that lies to the west, north, and south of Mossy Grove, rise several hundred feet above the community. High-velocity winds are rare in Mossy Grove. The uplands to the west and north shelter it from most of them. Also, the tornado occurred in the dark of night. Many people were attending night services at church or getting ready for bed and a new week of work or school when the tornado struck. Fortunately, the National Weather Service had issued a tornado watch and warning, so most residents were aware of the danger.

Without a doubt, the staff at the National Weather Service saved dozens, if not hundreds, of lives that night. The National Weather Service office in Morristown, Tennessee, issued a severe thunderstorm warning for Morgan and Scott counties at 6:18 p.m.. At 6:20 p.m., radar reflectivity images revealed the juvenile stages of a hook echo, which is a signature of a tornado. A hook echo on the back edge of a storm suggests that rain is being wrapped around a vortex. Storm-relative velocity images at that time also indicated rotation. Storm-relative velocity imagery shows the velocity of the wind minus the motion of the storm. When these two methods produce certain patterns, it can be inferred that there is rotation and, hence, a vortex.[41] The National Weather Service issued a tornado warning at 6:26 p.m. In just eight minutes, the warning rose from a severe thunderstorm to a tornado. At around eight thirty that night, an F3 with winds estimated to be 175 mph touched down west of Mossy Grove. It travelled east-northeast for 8.3 miles and dissipated near Joyner, Tennessee. Sadly, four of the victims were killed as they tried to escape from its path in an automobile.[42] Apparently, the driver was blinded by the intense rainfall and did not see the vortex, or it simply caught up with them. The supercell then continued eastward, and a new F2 vortex touched down at around nine o'clock in Frost Bottom, a quiet, narrow valley near the town of Briceville.

In its wake, the F3 left seven residents dead and an uncertain number of people injured. According to local residents Sarah Vann and Steven Williamson, an unrealistically large number of people claimed injuries from the twister. Apparently the lure of publicity via news outlets led some people to make unsubstantiated claims of being tossed about or otherwise injured by winds and debris. At the time, Williamson was ten years old. On that Sunday evening, he was visiting his grandparents living in Mossy Grove. According to Williamson, the tornado tore a path of destruction through the town about one hundred yards from his grandparents' home. The young man recalled that the lights in his grandparents' home were flickering on and off as the sounds of heavy rain and then damaging hail pelted the home's metal roof. Although there are no train tracks running through the small town of Mossy Grove, he heard a loud train in the distance. It sounded as if it were coming "down the mountain and swooped behind Mossy Grove along Jones Road. That's where the damage was most severe. It seemed like I could hear the train for a few minutes."[43] Among the survivors was a twelve-year-old boy named Quinton Woody. He was home alone and was apparently taking a shower when the tornado approached his Jones Road home. As the boy exited the shower, the tornado tore through his house and tossed him into a tree, breaking his left ankle and right forearm.[44]

While residents of Mossy Grove and Briceville were contending with their encounter with tornadoes, the National Weather Service office in Calera, Alabama, had issued numerous tornado alerts for eleven hours leading up to the outbreak. In Van Wert, Ohio, tornado sirens sounded fifteen minutes before the first of four vortexes ripped through the small city and the surrounding countryside. The Van Wert outbreak produced winds in excess of two hundred mph, achieving the destructive rating of an F4. Broadcasted warnings and tornado sirens prompted an alert theater manager to send about fifty matinee movie goers to interior restrooms. The manager's prompt actions saved dozens of lives because at 3:25 p.m., a powerful tornado hurled three cars into the front row of seats while ripping apart the exterior portions of the theater complex.[45]

SUMMARY

Tornadoes have captured the imagination if not fears of most Americans. From early cinematic wonders like *The Wizard of Oz* to the 1996 blockbuster *Twister*, Americans who have not experienced a tornado firsthand have at least seen one on the movie screen or through their television sets. Documentaries that feature storm chasers in hot pursuit of the elusive force of nature have been run on a number of cable channels. One program featured a man's efforts to film a tornado from the supposedly safe confines of a tornado-proof car. While programs such as this one promise to discover and present scientific knowledge about one of the most destructive forces on earth, they offer little more than visual stimuli for adrenaline-seeking viewers. Television programs often present images of storm chasers traveling across Tornado Alley. However, the chasers usually end up scratching their heads as they fail to film the big one. Tornado documentaries along with films such as *Twister* leave viewers with a sense that tornadoes are relegated to Oklahoma, Kansas, and maybe Texas. While the frequency of tornadoes is greater in the Great Plains states, the probability of a person falling victim to one is much greater in the southeastern United States and Mississippi River watershed, especially in the eastern portion.

In 1925, the Tri-State Tornado was born west of the Mississippi in the Ozark uplands. That storm killed 695 people but of that number, only thirteen were killed west of the river. The bulk of the victims were killed in Illinois. Of the top-ten deadliest tornadoes in American history, only one of them was located in Tornado Alley. That is not to say that life in the Great Plains is without risks from killer vortexes. The difference in death tolls is simply a function of numbers. There are a lot more people living east of the Mississippi than there are in the lands to the west. However, as population densities increase in the mostly rural areas in which tornadoes most often touch down, the death toll will likewise climb. There is little promise that homes can be built to withstand the 175 to 200 mph winds associated with an EF3 or an EF5. Such was the case in Mossy Grove, Tennessee, and Van Wert, Ohio, on November 11, 2002. Virtually nothing built aboveground by human hands can withstand the power of an EF5. Winds are only part of the destructive power of tornadoes; the objects they hurl are the real killers.

When the winds of the Tri-State Tornado subsided, rescuers working in the farmland near Parrish, Illinois, found the body of William Rainey in a stand of trees more than one mile from his home. One of his arms had been pulled off and the rest of his body was broken and lacerated. Tornadoes are not playthings to be chased and marveled at from roadsides. They can move at surprisingly fast speeds. The Tri-State vortex exceeded sixty mph for most of its 219-mile path of destruction. It also shifted directions. As the outbreak of April 27, 2011, shows, we should expect that at some time in the future, an EF5 producing three hundred mph winds will form over a farm, town, or city. Cars, trucks, timbers, bricks, and bodies will be cast about like leaves on a gusty autumn day. The only truly safe place to be in its path is underground. Perhaps more attention should be given to building underground shelters and installing sirens like those that saved lives in Van Wert, Ohio.

Chapter 9

LIFE AND DEATH IN QUAKE ZONES

INTRODUCTION

T he earth's crust is quite thin in relation to the size of the entire planet. Whereas the earth is eight thousand miles in diameter, its crust or the upper part of the lithosphere is no more than twenty-five miles thick under the continents and even thinner under the oceans. The lithosphere, which reaches a depth of about ninety-five miles, floats on top of a thin layer of plastic material called the asthenosphere.[1] Below the asthenosphere is sturdy mantle rock. Being brittle, the lithosphere can suffer breaks or faults as a result of stresses and friction caused by its floating on the plastic material below it. Earthquakes occur when energy is released by the movement of crustal or tectonic plates along or between faults. While earthquakes are not as predictable as hurricanes and tornadoes, scientists know a great deal about where tremors and destructive quakes will occur in the future. As was shown in the two previous chapters, there are regional "alleys" in which destructive atmospheric conditions develop. In the same general way, earthquakes can occur in predictable places.

The most common or likely place for earthquakes to occur is along the Pacific Rim, an area known as the Ring of Fire. This part of the earth's surface is also marked by volcanism, which is a process of mountain building resulting from the emission of molten lava through fissures in crustal plates. The reason for the high frequency of seismic activity there is because this part of the earth's lithosphere features a series of subduction zones where friction generated by opposing plates causes fractures and fault zones. In the

Southern Hemisphere, there are a number of plates involved in lithospheric movements. Because Australia sits in the middle of its own plate, it is one of the safest places to live in regard to earthquakes and volcanoes. Australia's wide variety of mammal-consuming reptiles, poisonous spiders, and venomous snakes, however, are another matter. In the Northern Hemisphere, knowledge of seafloor spreading, especially in the middle of the Atlantic, has added to our awareness of subduction zones that form on the Pacific Rim. As North America is pushed westward, Eurasia is pushed eastward. The North American and Eurasian plates are, hence, "pushed against" the thinner Pacific Plate. The Pacific Plate is thrust under (or subducted by) those continental plates. These subduction zones create stresses on the lithosphere that create earthquakes and volcanic eruptions. Arguably the best-known fault that has resulted from this process is the San Andreas zone that lies under Mexico and California. Numerous tremors are detected along the zone every year, but most of them are not felt by residents. Such was not the case in 1906 when a major quake struck the Bay Area of San Francisco.

Volcanoes are common where heat-generating friction causes rocks to be converted into molten material. Magma is either pushed through fissures where it is continually emitted as lava, as is the case with Kilauea in Hawaii,

Image 9.1. San Francisco with the bay in the distance. *Image courtesy of the author.*

or it is periodically spewed through violent eruptions such as the one that occurred at Mount Saint Helens on May 18, 1980. As can be seen in chapters 7 and 10, volcanoes also emit varying quantities of ash, which, when present in the atmosphere, has the capacity to reflect sunlight back into space. Major eruptions occurring in tropical regions have the greatest impact on global temperatures because it is in this latitudinal zone (40° north to 40° south) that surplus energy is absorbed into atmospheric and oceanic systems and then dispersed poleward into zones of energy deficits. Volcanic eruptions along the Pacific Rim can have serious impacts on global weather patterns for as long as one year. When combined with earthquakes, volcanoes make those coastal areas dangerous places to live. However, one of the deadliest earthquakes to strike terror in the hearts of people did not occur along the Ring of Fire.

The deadliest earthquake in centuries, if not in history, occurred on December 26, 2004. The primary cause of death was not the result of collapsing houses and apartment buildings, as one might expect. Instead, this earthquake's epicenter was along a fault zone located under the Indian Ocean. Rock on one side of the opposing plates was moved vertically several yards. This uplift occurred along a six-hundred-mile stretch of seafloor. When the uplift occurred, energy equal to twenty-three thousand Hiroshima-type bombs was unleashed. A pulse of energy that moved as fast as a jet airliner spread out toward India, Myanmar, Bangladesh, Indonesia, and the east coast of Africa. In the open sea, the pulse created a wave that was a little more than one foot in height, but as the wave moved into shallower water, friction along the seafloor made the upper layers of water pitch forward, creating waves as high as fifty feet. However, in most cases, the water coming ashore simply rose quickly and pushed inland as if it were a high tide on steroids. The tsunami that resulted from the Indian Ocean earthquake killed an estimated 150,000 people.[2] Many of them, even those in posh resorts, did not have any idea the earthquake had occurred, nor did they know that a tsunami was headed their way.

Another major seafloor earthquake struck on March 11, 2011. Its epicenter was about eighty miles east of Honshū Island, Japan's mainland. With a Richter Scale rating of 9.0, this particular quake was the strongest ever recorded in Japan and the fifth strongest on record anywhere in the world.[3] Eastern Japan has some thirteen hundred miles of coastline, and dozens of cities were shaken by violent tremors and, sadly, others were destroyed. Twelve days later, police estimated that the death toll had exceeded eighteen thousand.[4] The closest city to the epicenter is Sendai, where local officials reported that a twenty-three-foot-high tsunami inundated the city. When the waters receded, more than three hundred bodies were found near the coast.[5] In Ofunato City, three hundred homes were washed away by the invading Pacific. Because a nuclear power facility and an oil refinery had been damaged, it was necessary to evacuate thousands of people. In Tokyo, which is located about 240 miles from the epicenter, four million buildings were without power, and high-rise buildings shook for more than two minutes, leaving cracks in walls and equipment scattered around normally tidy offices. Millions of workers ran down stairwells and entered streets and highways to escape falling debris. What is amazing about this quake is that the loss of life was not greater. Japanese construction engineers and building companies, realizing their precarious location along the Ring of Fire, built high-rise structures that were able to absorb heavy shaking. Thanks in large measure to tsunami warnings and advisories, millions of people living along the Pacific Rim were warned and were able to evacuate coastal areas.[6] Across the world's largest ocean, on the coasts of Northern California, minor wave action was detected. Higher waves were observed in the Hawaiian Islands.

The massive number of deaths resulting from the 2004 quake, however, was in part caused by the high population density along the coastline of the Indian Ocean. This portion of the ancient ecumene was settled by some of the earliest emigrants from the Near East. Because these seemingly benign shorelines have historically provided people with an abundance of natural resources, the carrying capacity is high, but the unpredictable nature of earthquakes and the human geography of the area combined to make December 26, 2004, one of the deadliest days in human history.

In their efforts to help societies cope with these formidable forces of nature, geologists, geographers, and other volcanologists have worked for decades to develop systems to predict the occurrence of earthquakes and volcanic eruptions. Their work has mostly concentrated on identifying patterns of vibrations that they can detect on seismographic instruments. Construction engineers have also developed "flexible" building designs that allow structures to move in response to shaking earth. To date, however, no one has designed a building that can resist lava flows from volcanic eruptions. Being aware of the major earthquakes that shook San Francisco in 1906 and again in 1989, as well as the Northridge quake of January 17, 1994, the geographic imagination of earthquake alley among Americans is understandably focused on the Pacific coast. With images of the eruptions of Mount Saint Helens in 1980 burned into the memory of those old enough to remember it, there is certainly little to make Americans appreciate the fact that arguably the most powerful earthquake to strike American soil did not occur in California or Washington. "On the basis of the large area of damage (600,000 square kilometers), the widespread area of perceptibility (5,000,000 square kilometers), and the complex physiographic changes that occurred, the New Madrid earthquakes of 1811–1812 rank as some of the largest in the United States since its settlement by Europeans."[7] The reason that no one can state for certain that the New Madrid quakes, whose epicenters were along the Mississippi River in northeastern Arkansas and southeastern Missouri, were the largest in American history is because there were no seismographs at the time to record objective data on the quakes, let alone on their magnitudes. That capability did not happen until 1935, when Charles Richter and Beno Gutenberg of the California Institute of Technology developed a scale that they intended to be used locally in Southern California. Their method used a 10-base logarithm scale to measure shaking or horizontal amplitude of local magnitude.[8] Nevertheless, given that the strong area of shaking caused by the New Madrid quakes were ten times as large as that which devastated San Francisco in 1906, a strong argument can be made for its place in geographic history, or as H. C. Darby would call it, "geography behind history."[9] Before presenting a historical geography on that major event in America's environmental past as well as what impacts might result if another "big one" were to hit that region, it is important to consider how earthquakes are measured and how predictions are made. As we will see shortly, this is an evolving science—one that has a long way to go before it reaches the same level of accuracy that we have come to expect from the National Weather Service when it predicts hurricanes and tornadoes.

PREDICTING THE NEXT BIG ONE

Predicting the next big one is not easy, nor is such a practice currently reliable. However, thanks in large measure to seismologists who first theorized primary, secondary, and surface waves in the early 1800s, identifying places where earthquakes will occur is indeed reliable. Just as early leaders of the National Weather Service argued that in the 1880s the science behind predicting tornadoes was inadequate and would unnerve citizens, the same can be said for making predictions about when a major earthquake will occur.

As a young, aspiring geoscientist, I can personally attest to the visceral reaction some people may have to the fear of an earthquake. In 1989, I was a graduate student immersed in the geosciences at Western Kentucky University. That fine institution is located about sixty miles north of Nashville, Tennessee, in Bowling Green, Kentucky, which is situated in an area that since 1985 had been predicted to experience widespread damage from an eminent earthquake.[10] The epicenter of this anticipated quake would be along the Mississippi River, in the same location as the big ones that occurred there in 1811–1812. While my reaction was not a major event in and of itself, the fact that I and other graduate students were assigned the task of mapping caves in the damage zone led to a situation that affected the mapping of a portion of the regional landscape, albeit on a minor scale.

Professor Nick Crawford, the university's well-regarded hydrologist, asked Mike Brent, John Wilson, and me to use a topographic map to locate and map a series of caves in Simpson County, Kentucky, that Crawford had theorized were connected to the Mammoth Cave system. In exchange for our work in being the first persons to map these caves, we were allowed to name portions of them. Armed with a host of mapping tools, we entered Wilson's Cave on a cold December night. As we descended into the "bowels" of Simpson County, we encountered a shallow pass that connected two larger rooms. I was the second of the group of three to crawl through the thirty- to forty-foot section. The ceiling of this corridor was only twenty inches above its floor, which made me feel as though I was in a vice grip. When I was about half way through, scooting on my back, face up, thoughts of the recent earthquake predictions suddenly took over my mind. In my imagined world of the New Madrid earthquake, I watched the wave emanate from the epicenter along the Mississippi River. Then I imagined waves of energy passing through the limestone rocks in which I quickly perceived myself to be sandwiched. Within seconds, I imagined the ceiling above me collapsing, leaving 225 pounds of gooey stain between two layers of Simpson County limestone. The stain, of course, would have been me. My reaction was perhaps one of the most extreme panic attacks my friends had ever witnessed. With their help, I calmed down, and we finished our chores. Once back on campus, we sat down to name rooms and portions of Wilson's Cave. Those names were to appear on the maps we had drawn for the Center for Cave and Karst Studies.[11] My colleagues decided that the shallow, narrow pass that prompted me to imagine my demise would henceforth be known as "Panic Pass." It has been over twenty years since we mapped Wilson's Cave and, thankfully, another big one has not visited the region. Nevertheless, this situation shows the visceral reaction some of us can have when we surrender to the fear of an earthquake. The lack of precision in making predictions can invite unnecessary or unfounded worry. With that said, however, the continuance of minor tremors in the New Madrid zone and the fact that the region was visited by big ones before make it necessary for structures to be engineered with earthquakes in mind.

While predicting the next big one with current seismographic instruments has limited merit, current technology can show the location of quake epicenters. Knowing the geography of particular earthquakes allows rescue teams to be sent out quickly to areas that were hit the hardest. It also lets us know where the next big one might occur. The ability to map earthquakes has evolved a great deal since the early 1800s, when Augustin Cauchy, Siméon Poisson, and others began the process with their work on elastic wave propagation in the early 1800s.[12] A brief look at how earthquakes are measured deserves some attention because it informs our later discussion on the geographical history of some of America's greatest quakes.

The person to coin the terms "seismology" and "epicentre" (Americanized to "epicenter") was Robert Mallet (1810–1881), an Irish engineer. In 1849 he and his son John, who, at the time, was a geology student at Trinity College Dublin, conducted a series of experiments on the east coast of Ireland to determine if waves of energy emanating from an epicenter could travel through sand. Detonating dynamite buried in the sand, the father-son team used a seismoscope to detect waves of energy one-half mile away from the explosive epicenter. In 1857, a large earthquake struck southern Italy, and with the help of Charles Darwin and English geologist Charles Lyle, Mallet received a grant from the Royal Society of London to travel to Italy so he could study the area. His field trip enabled him to draw maps illustrating the contours of damage.[13] Mallet's research and maps suggested that energy from earthquakes radiates outward from an area of focus, the epicenter. It also showed that locating the epicenter of an earthquake is possible by tracing damage waves backward, toward the source. Mallet was convinced that widely established monitoring stations would yield important data that would help build earthquake science. Later that year, Mallet and his son drew one of the first seismographic maps of the world.[14]

It should not come as a surprise that the next advances in the development of seismographs would occur in societies that had the most to gain from innovations and refinements in the fledgling science. Filippo Cecchi (1822–1887), a physicist from Tuscany, was barely thirty-five years old when the quake of 1857 struck his native Italy. Being a man of many talents and interests, Cecchi was also keenly interested in meteorology, and in 1885, he was elected vice president of the newly formed Italian Meteorological Society. Whereas in the United States there is a clear professional line between meteorology and seismology, Italy had no such distinction in the late 1800s. Cecchi directed the Osservatorio Ximeniano in Florence from 1872 to 1887, and he helped establish other meteorological stations in Tuscany. Using his own funds, Cecchi set up an endowment for his Florentine meteorological observatory to operate a seismological center.[15] Although Cecchi invented various types of seismographs, he is perhaps best known for the one he designed and built in 1875.[16] While his instrument was reliable in measuring waves in the early stages of an earthquake, its accuracy diminished after the first few cycles of shaking. The Cecchi seismograph was put to good use in other countries. F. Du Bois published a description of the Cecchi instrument that was installed in Manila in the Philippines in 1885.[17]

Until 1897, seismometers, including the Cecchi device, were limited in their ability to chart successive waves of energy produced by an earthquake. That all changed when E. Wiechert developed a widely used seismometer that was able to record the entire duration of a quake.[18] In the same year, a seismometer was installed at the Lick Observatory near San Jose, California. This was fortuitous because the instrument was still in use when the San Francisco earthquake occurred in 1906 and the Wiechert seismometer recorded valuable information about that devastating event. An American engineer named H. F. Reid studied survey lines across the San Andreas Fault before and after the San Francisco quake and developed an elastic rebound theory, which proposed that accumulated elastic energy was built up along the fault. The quake occurred when the energy was released. Subsequent research has confirmed that the rebound theory is the principle cause of tectonic earthquakes.[19]

The Wiechert device, however, was not the last major step forward in monitoring earthquakes. The first of the electromagnetic seismometers, which are still used today, was developed in Russia by B. B. Galitzen and were installed in a number of countries after WWI.[20]

While detecting and measuring earthquakes has enabled seismologists to identify places where the next big one will likely occur, it is difficult to know how much stress is required to trigger it. One fact is certain: earthquakes originate in fault zones. It follows that places that have been visited by tremors and massive quakes in the past are likely to experience them again. Unlike a broken femur or tibia, fractures in the lithosphere do not heal. As described by Reid in his rebound theory, tectonic forces that build stress between plates do not disappear when the tension is released in a quake. Due to seafloor spreading, the lithosphere persistently reloads forces that cause tectonic activity and earthquakes. While past experiences with quakes are important in constructing probability models to predict the next big one, being able to measure plate stresses and how much pressure is needed to trigger an event are elusive challenges.

MEASURING QUAKES

While there is no clear way to predict the time when an earthquake will occur, seismologists have developed a scale to measure their power. Just as hurricanes are known by their category ratings and tornadoes are described by their F-Scale placements, earthquakes are measured on a Richter magnitude scale. Developed in 1935 by Charles F. Richter of the California Institute of Technology, the Richter scale measures the amplitude of energy waves generated by earthquakes. The scale takes into account the distance of the various seismometers that measured the quake and its epicenter. Unlike the Fujita Scale for tornadoes, the Richter method does not consider damage to infrastructure. Waves are measured by whole numbers with decimals (i.e., 3.2, 5.4, etc.). The scale can only be used with seismometers that are calibrated to its specifications. Since the scale is based on logarithms, each whole number represents a ten-fold increase in the amplitude of a wave. To put that thought into more imaginable terms, each whole number represents thirty-one times more energy.[21]

The earth's lithosphere is quite active, and earthquakes remind us of that fact. Quakes registering 2.0 or less, which are not felt be people, are called microearthquakes. Moving into the range of shaking that is perceptible to humans, earthquakes of at least 4.5 on the Richter scale are common. Several thousand of them occur each year. Although the scale does not have an upper limit, great earthquakes fall in the 8.0 to 9.0 range. On average, there is one great earthquake somewhere in the world per year.[22] For comparison purposes, the United States Geological Survey (USGS) considers a quake registering in the range of 5.3 to be moderate, and one that registers 6.3 is labeled as strong. It is important to point out that these are measures of energy as expressed in waves recorded on seismographs. In more practical terms, if a moderate quake were to occur in an area in which buildings are constructed in a rigid fashion with little ability to move in response to the shaking earth beneath them, the human death toll could be high. On the other hand, a strong earthquake could strike an outlying forested area and the human death count could be light.

While the Richter scale is a measure of energy generated by an earthquake, its meaning is not easily understood by the average person on the street. In a manner that Ted Fujita would later use to measure and categorize damage caused by tornadoes, in 1931 Harry Wood and Frank Neumann developed a scale for describing damage caused by an earthquake.[23] It is generally understood that the tangible elements of human societies are part of the landscape, and as such, the effect of a quake on the landscape or surface of the earth is called intensity. Wood and Neumann kept those thoughts in mind when they developed their twelve-stage scale called the Modified Mercalli Intensity Scale. Apparently to give their taxonomy something of a classic look, they employed roman numerals in creating their categories. Nevertheless, the lower numerals reflect human perceptions of the event, and the larger numerals identify various levels of damage to infrastructure.

In essence, the twelve stages describe intensities that range from "imperceptible shaking" to "catastrophic damage." Structural engineers are called upon to assist in assigning Mercalli intensity values of VIII or higher. A Mercalli I quake is not felt by people, a level II tremor is felt by a few people at rest in the upper floors of buildings where base-generated shaking is more pronounced. A quake classified as a level VII event breaks chimneys while causing considerable damage to buildings of poor design. Well-built, flexible structures, on the other hand, only experience slight damage. A level X quake destroys some well-built wooden structures, but most masonry buildings are destroyed. Even rail lines are bent by their powerful undulations. An earthquake at the top of the scale produces total damage. Objects can be thrown in the air because of the severe movement and shaking. The Modified Mercalli Scale is certainly helpful in comparing the effects of earthquakes, but it does little to warn people of the likelihood of a tectonic event.

Societies located along the Pacific Rim, the area of the earth that is perhaps most likely to experience strong to great earthquakes, are cognizant of their situations. In recent years, building designs have been implemented that enable structures to absorb most of the shock generated by tectonic activity. Still, a great quake will be difficult to withstand. Settlements located on the shores of the Mediterranean Sea are vulnerable because a large number of their buildings are old and brittle. Fortunately for those along the Mediterranean Sea, earthquakes are less frequent, but they can and will occur again. Given the high population densities of Greece and Italy, the death toll could be high. As an area of active volcanism, those societies face a second dangerous scenario from plate tectonics. It is not pleasant to imagine the devastation and loss of life that would occur in the Mediterranean Basin if it experiences eruptions like those that occurred at Santorini in 1500 BCE and at Mount Vesuvius in 79 CE.

NEW MADRID EARTHQUAKES OF 1811 AND 1812

Aside from the Pacific coastal areas of California and Mexico, the New Madrid fault zone in North America is expected to experience another strong earthquake in the relatively near future, but the public's preparedness is questionable. As discussed earlier, the awareness of the dangers along the Mississippi River are known by seismologists. However, because of the wide expanse of time between major events as well as a lack of precision in making predictions, future scenarios invite something of a dismissive attitude among members of the public, unless, of course, they find themselves spelunking in the predicted damage zone between two tight layers of limestone rock about two hundred feet below the earth's surface. It is easy for laypeople to lose sight of the possibility that a quake in the range of X to XII is looming on the temporal horizon. In 2003 the USGS revised the 1985 forecast that caused my subterranean panic attack. The recent forecast projects a 7 to 10 percent probability that an earthquake in the Richter scale range of 7.5 to 8.0 will occur in the New Madrid fault zone sometime in the next fifty years. The probability is quadrupled for a quake in the moderate to strong range.[24] Clearly, the fault lines that produced the 1811 and 1812 quakes have not healed themselves. Tension still exists among and along those tectonic fractures.

Most of the authors who have written on topics related to the three main earthquakes that occurred in this area two hundred years ago owe a great deal of gratitude to the considerable effort and reconstructed model of Otto W. Nuttli (1927–1988), a professor of geophysics who worked at St. Louis University.[25] Nuttli used period newspaper accounts that detailed the sequence of earthquake events to draw an isoseismal map of the first of three principal shocks. He also noted that there were upward of two hundred aftershocks, and remarkably most of them were in the moderate to strong range. Nuttli concentrated on the three largest that probably registered 7.2, 7.1, and 7.4 on the Richter scale, respectively. His map reflects an unusually large-felt area that extended to the Atlantic coastal area of the Southeast. His examination of old newspaper accounts apparently did not yield many stories about damage to the west of the epicenter. Given the time period and the absence of newspapers in the few settlements located west of the Mississippi, he had little to study.

This was a time when pioneers like Daniel Boone (1734–1820) were hunting the eastern forests of Missouri. It was also a time in which Tecumseh and other leaders of Shawnee, Kickapoo, and Winnebago tribes sought intertribal unity to resist the ever encroaching advance of white Americans into their lands. Few observations of members of native tribes that lived in the area were recorded for posterity, so their experiences and reactions to the earthquakes were not well documented. Still, in 1811 and 1812 the number of Native Americans and Europeans living in the areas immediately west of the Mississippi was small in comparison to the population density today.

The two largest settlements between New Orleans and St. Louis were New Madrid, Missouri, and Natchez, Mississippi. Because the quakes more directly impacted New Madrid, its name was affixed to the sequence of quakes that rocked the area. Whereas primary (P) and secondary (S) waves ripple through deep bedrock at a high velocity, surface energy such as Rayleigh and Love waves move much slower. They were the kinds of movements that leveled New Madrid, Missouri.[26]

Those surface waves create two distinct patterns of motion that, when combined, create more instability for any structures that sit astride them. In a manner similar to swells that move along the surface of an ocean, Rayleigh waves create up and down elliptical surface movements. The shaking of a building hence resembles a cork bobbing up and down, toppling backward and forward throughout the event. On the other hand, a Love wave causes side-to-side motion. It would take an exceptionally well-designed building with structural flexibility to simultaneously withstand these opposing forces of motion. In 1811, such was unheard of along the western frontier.

The New Madrid earthquakes were among the most intense tectonic events in the natural history of North America. East of the Rockies, they were certainly the strongest. The area of strong shaking was ten times the size of that which was affected by the 1906 San Francisco quake. In addition, physiographic features near the epicenter were also altered. Reelfoot, the only natural lake in Tennessee, was formed by the third event. The first earthquake studied by Nuttli occurred in northeastern Arkansas at just past two o'clock in the morning on December 16, 1811. It was followed by a large aftershock at 7:15 a.m. Based on firsthand accounts, Nuttli decided that this was the largest of the two hundred or so aftershocks. It may well have registered with a magnitude of 6.5. On January 23, 1812, the second major earthquake reverberated from an epicenter in southeast Missouri. Finally, the third quake occurred along the Reelfoot fault in northwest Tennessee and southeast Missouri on February 12, 1812.[27] The area that experienced level VII damage on the Mercalli scale was large and impressive. Based on accounts of witnesses, the first quake caused structural damage throughout an area 600,000 square kilometers (231,600 square miles) in size. To put this into perspective, the area of structural damage was twice the size of Arkansas and Missouri. It is reasonably well established that the public was alarmed (Mercalli level V) throughout an area spanning 2.5 million square kilometers or 965,000 square miles.

These quakes caused surface areas of the region to rise and fall. Islands in the Mississippi River disappeared and shifts in the riverbed generated waves that travelled north of the epicenter, creating the illusion that the "Mighty Mississippi" was forced to flow backward. Cracks appeared in the landscape and landslides were observed along river bluffs and hills. In areas of subsidence, aquifers were broken and new springs were formed. The most obvious impact that occurred on the landscape was the formation of Reelfoot Lake in Tennessee where subsidence was between 1.5 (4.92 feet) to 6 meters (19.68 feet). As a result, the lake that now sits in the depression is quite shallow.

While the impact on the landscape was severe, the fact that few towns were located in the immediate area meant fewer deaths. Only one death was caused by collapsing structures in New Madrid. However, cabins and chimneys were thrown down as far away as Cincinnati, Ohio, just under four hundred miles away. Similar structural devastation occurred in Kentucky and Tennessee.

SUMMARY

Earthquakes and volcanoes have always plagued humanity. Perspectives on and responses to this force of nature have progressed from geotheokolasis beliefs to knowledge of energy waves as well as accurate maps depicting seismic activity and fault zones. We know where deadly tectonic events have occurred and where they are likely to happen in the future. Forecasting future events, however, is fraught with ambiguities because tension exists along hundreds of fault lines, and as was seen in the Indian Ocean tsunami of 2004 and the New Madrid earthquakes of 1811 and 1812, they can occur in places that the general public does not expect. As is the case with images of Tornado Alley being situated exclusively in the Great Plains, imagery of earthquakes seems to be focused on California and the Pacific Rim, although thanks to Abrahamic texts, archeology, and observations recorded by Greek and Roman philosophers, tectonic forces around the Mediterranean Sea are also well known. Places like the southern Caribbean Basin, which we have not discussed, the Indian Ocean Basin, and the Mississippi River Valley are not on the "seismic radar screens" of many people outside of the circle of scientists who study and write about such things.

If the forecasts issued by the USGS are accurate, the future does not bode well for people living along the Mississippi River or its major tributaries. With the rise of cities like Memphis, Tennessee; West Memphis and

Jonesboro, Arkansas; Cape Girardeau and St. Louis, Missouri; as well as Paducah and Owensboro, Kentucky, there is no precise way to accurately predict the casualty rate that would result from a Mercalli IX or X event. There is a great deal of variability in building designs and construction materials used in the region. Guesses can be made, however, but, as a former spelunker, I know firsthand that dire predictions can create a good bit of anxiety and panic under the right circumstances. In the final analysis, however, some anxiety can be a good thing if it causes people to be aware of the types of structures they build and occupy.

Chapter 10

CHALLENGES OF CLIMATE CHANGE

INTRODUCTION

As it was among the ancients, public discourse involving weather and climate often used emotionally charged, fearful terms and phrases like "acts of God," "wrath of God," "catastrophe," "danger," "extinction," "collapse," and "terror." Throughout human history and even today, ideas to lessen fears associated with adverse weather have been accompanied by calls for repentance or changes in behaviors. In recent years, proposed attempts to lessen those fears often require the regulation of individuals as well as control and mastery over the planet.[1] The civic dialogue certainly drives or is driven by international economic and political forces. This is seen in the mission of the Intergovernmental Panel on Climate Change (IPCC). The panel was created in 1988 and operates under the auspices of the United Nations. The mission of the IPCC shows a preoccupation with human-caused climate change while dismissing the "what ifs" of natural climate change. The panel seeks "to assess on a comprehensive, objective, open and transparent basis the scientific, technical and socio-economic information relevant to understanding the scientific basis of risk of human-induced climate change, its potential impacts and options for adaptation and mitigation. IPCC reports should be neutral with respect to policy, although they may need to deal objectively with scientific, technical and socio-economic factors relevant to the application of particular policies."[2] To the IPCC and many others, there is a genuine fear of what we are doing to the planet.

As a geography professor, I have listened to worried students express the belief that "the earth is in danger. We must stop killing the planet." To feel as though they are helping to heal the planet, some of them have become vegans. They have convinced themselves that they can live their lives without exploiting natural resources and that they can erase their "carbon footprints." That is a noble and widely held sentiment, but, in practical terms as well as in the scope of the earth's climatic history, nothing could be further from the truth: in the absence of CO_2 and adequate atmospheric moisture, an ice age will return with deleterious consequences for most of the economically developed world and the plants and animals that dwell therein. We are not importing CO_2 from other planets, nor are we manufacturing it. The earth has a fixed amount of carbon and it is recycled through natural processes. Unfortunately for us, nature has a way of annually storing small amounts of it in limestone rocks, chalk beds, and coral reefs. If we released all of the stored carbon back into the carbon cycle, would not the climate then resemble that of the Carboniferous age? Indeed, nature's way of storing CO_2 and H_2O over time may have actually contributed to the ice ages that have dominated earth's climatic history for more than one-half million years.

Beyond that substantial point, which is at the heart of this chapter, the idea that my earnest students can live their lives without exploiting resources is virtually impossible. The clothes they wear, for example, are made from materials that were once alive or otherwise resting peacefully somewhere. Even cotton was once alive, and the plants on which those fibers grew were subjected to defoliants to speed up the harvesting process. The rice-, oat-, or wheat-based cereal the students ate for breakfast was certainly once alive and grown for the express purpose of feeding them and millions of other people. The farm equipment used to cultivate these crops depends heavily on carbon-based fuels. Oh, yes, the computers and televisions they use to watch *Dancing with the Stars* and *American Idol* are made out of plastics refined from petroleum-based carbon, which once strode the earth as *Tyrannosaurus rex* and other dinosaurs. The labor used to assemble those seemingly nonexploitive tools for human enjoyment was supplied by workers making meager incomes. Even the act of breathing, which produces CO_2 as a byproduct, is an act of resource exploitation.

Companies and marketing agencies have noticed that people have an admirable desire to live in harmony with nature. They have responded with marketing campaigns that suggest that by using their products, consumers are helping to heal a hurting planet. No matter how clean or efficient a product may be, its manufacture and use are nevertheless acts of exploitation. "Well," one might think, "planet earth would be better off if we humans weren't here to exploit her." That notion ignores the fact that the earth's environment allows for some exploitation. Indeed, it demands it. The goal to stop "exploiting her" applies a gender label onto an object (earth) that is not alive, suggesting that humans project themselves onto all sorts of things. We seem to apply the Golden Rule to plants and nonliving objects that have no awareness of anything, including us or our benevolent attitudes toward them.

In their own self-interests, all plants and organisms manipulate and transform resources without any conscious thought as to the impact they have on other members of their respective biological communities. The body of a dead human is nothing more than a collection of natural elements that decomposers will exploit for their benefit. The earth's systems and elemental cycles are full of conflicts and biotic exploitations. It is too bad that Simba and Mufasa never fully explained the "circle of life" in the 1994 smash hit *The Lion King*, for if those imaginary characters had, my first-year geography students would know that life, whether plant or animal, depends on the magnificent manner in which life reproduces itself through the cold act of nutrient recycling through interspecies exploitation.[3] Carbon-based energy to propel limbs, lungs, and gills is passed through lowly algae and grass through animals such as gophers and lake trout to decomposing microorganisms where the entire system of resource exploitation begins anew. However, as mentioned above, a small amount of carbon is annually removed from the elemental cycle and stored in sea coral and under sediments where it hardens or lithifies into limestone and chalk.

Energy moves through a real circle of life called the food chain, albeit under the control of the two laws of thermodynamics. Those laws tell us that although energy cannot be created or destroyed, it can be converted or transformed. When energy is transformed, some of it is lost as heat. For decades, this concept led biologists to divide life forms that make up the food chain into "producers" and "consumers" such as grass and cows, but that dichotomy leads us to view life-forms as either beneficial or as consuming exploiters; hence, it is inviting to believe that only photosynthetic plants have value. Upon closer inspection, however, each organism and plant, whether large or small, cannot exist alone. Even a plant needs nutrients from the soil. Without microorganisms to decompose or break down feces and urine as well as the bodies of animals and plants, the soil on which plants depend would become sterile. Plants would die and so would the animals that eat them.

The notion that humans are somehow killing the planet is absurd, although we can kill ourselves and some other fellow occupants. As apex predators, we have the capacity to manipulate earth's resources in ways that no other life-form can. While on the one hand playing with fire may cause the player to get burned, he nevertheless must burn energy to survive. Exploiting resources is part of life. It is how we use them that must be done with care because the earth is not fragile; we are. From looking at environmental history, one fact rises equally from the volcanic ashes of great eruptions and the slippery slope of advancing glaciers: climate and weather are products of a complex energy system that neither sees nor knows us. They have no consciousness or need for sacrifice on our part. They cannot be bought off by repentance. Moreover, climate is capricious and because it is that way, we face a future that is likely to change. Perhaps the greatest challenge in facing the real prospect of climate change, which will happen with or without our assistance, is to shed ourselves of overly romantic and even geopious notions of perceiving weather and climate as anything more than mindless forces of nature.

WHEN HUMAN ECOLOGY
APPROACHES GEOTHEOLOGY

The interplay between living and nonliving elements in the environment is called ecology, and, when the focus is specifically on humans, geographers get excited because human ecology is an important paradigm within the discipline. As Aldo Leopold reminded us in *A Sand County Almanac*, we must develop and maintain a land ethic that regards humans as part of a community of organisms that depend on each other for survival.[4] However, Leopold's interpretation of a land ethic is fraught with beliefs that are not compatible with the nature of *Homo sapiens* or other members of a biological community. For instance, he regards a healthy land ethic as one that "changes the role of *Homo sapiens* from conqueror of the land-community to plain member and citizen of it."[5] Leopold clearly had a negative view of the profit motive as a reason for conservation. To him, the real and lasting motive for conservation was much deeper, touching on an almost spiritual sense of one's place in the community of living and nonliving members of the ecological system. Upon closer examination, Leopold leaves behind the basic notion of a biological community and embraces a belief that humans can take part in a "psychological community" that demands reciprocated emotional bonding with other organisms and nonliving elements. While Leopold was an unabashed proponent of evolutionary biology, he nevertheless allowed his ideas about nonhuman members of the community to morph into a belief system resembling animism. Animism posits that both living and nonliving objects in nature have souls and are hence able to reciprocate in psychological bonding with humans.[6] Moreover, the notion of citizenship implies membership in a social organization that carries with it rights and responsibilities. Is it healthy to believe that a copperhead, a rattlesnake, or a deer tick sees mammals as anything other than a threat or perhaps a meal? A healthy land and climate ethic need not reach the stage where it becomes a religion or a political ideology, for when it does, the result on humanity could be as troubling as having no ecological sensitivity at all. At that stage, logic is cast aside in favor of geopious (the emotional side of religion tied to space and nature) sentiments, and nature, as was viewed by the ancients, is once again regarded as sacred or divine. Aside from some technological innovations, what else would then separate us from our ancestors who worshiped nature gods?

As we saw in chapter 4, even the most secular person among us is liable to see humanity as engaged in a punishment-reward relationship with nature, geokolasis and geomisthosis, respectively. In a similar manner, my colleague Mike Hulme notes that during the early modern period of European history, which would have included the last half of the most recent "little ice age," climate was seen as a means of judgment but today is seen as a form of social pathology.[7] Spiritual connections to the earth and its systems are much deeper than most of us might think. Émile Durkheim (1858–1917), a pioneering French sociologist, was a keen observer of religious practices and symbols. In his seminal work *The Elementary Forms of Religious Life*, which was originally published in 1912 and recently revised with the help of Carol Cosman and Mark Sydney Cladis, Durkheim describes how a number of societies have created religious totems that take the form of a wolf, bear, or some other creature that they feel represents their core values and essence as a people.[8] Their totems were generally taken from their local environment with which they felt a special kinship. It is helpful to see that the concern many people have for the preservation of the earth as a living being is perhaps a subconscious projection of the planet onto a totem matching the evolving identification many people have with global citizenship that is ultimately a product of global economic interdependence. Whereas in the past a tribe of Native Americans who lived in a restricted area of the Pacific Northwest selected an object from nature they felt epitomized who they were as a people, more globally minded folk today have selected the planet as their totem. Of course, they would not see their identification with the earth as a totem in such simplistic terms because they assume that we humans have socially evolved to a point where we are no longer dependent of simplistic religious notions. It is helpful to recall from chapters 4 and 9 that perhaps a majority of Americans 150 years ago would have subscribed to the notions of geotheokolasis and geotheomisthosis; today those beliefs among secular people have morphed into geokolasis and geomisthosis. While God may have been expunged from their belief system, they nevertheless still have a belief system that sees "mother earth" or "nature" as a living, breathing entity that rewards people for good stewardship while threatening to punish them for exploiting the good which she or it has to offer them. It takes but a little thought to see that the real object or totem of veneration is not the earth, but, as Durkheim would argue, it is us. It follows that for globally oriented people, the earth is endangered by our actions. However, just as the totem reflects some aspect of the society that created it, the planet in their worldview represents what can happen to us through our own actions.

STIGMATIZATION OF CARBON

To be sure, there are circles of life, and all biotic forms exist because there is an optimum combination of resources, gases, elements, and climatic conditions that are conducive to life as we know it. Physics professor Guillermo Gonzalez and philosopher Jay Wesley Richards effectively describe this remarkable situation in their book *The Privileged Planet*.[9] The earth's biosphere is an amazing place where ecological systems recycle elements and resources. Key among the thirty-five elements that support life is nitrogen and carbon, and both of them are recycled. Together with water, these elements make up the real circle of life. It is hard to believe that the glasses of water or water-containing beverages we consume were once part of the lifeblood of a grizzly bear, a porcupine, or even a mass murderer. As an innocent victim of the global warming debate, carbon, especially carbon dioxide, has become maligned. Carbon, including gaseous carbon dioxide, is nonetheless necessary for life, and, as such, it is a key element in all ecosystems. The more we know about the element and how vulnerable and how valuable it is to us, the easier we can adapt a healthy land and climate ethic.

The vitality of a land and climate ethic, however, is threatened by a growth in urban areas and a decline in rural settlements. Urbanization is arguably driving a wedge between modern people and an awareness of their reliance on food webs and exploitation of other natural resources. Sadly, to some folk, food just appears on grocery store shelves, so they fall victim to simple marketing ploys that take advantage of their lack of knowledge of the connectedness of the food web to themselves. To place that thought into context, would you rather buy "country-fresh" eggs and "down-home" bacon, or just eggs and bacon? If these items were taken from the chicken and hog last week, what is the difference between the two? The use of the word "country" in one choice suggests that those eggs are closer to the wholesome earth, but unless the packaged eggs and bacon were grown in the city, which they were not, marketing and labeling is the only difference between the two options.

In Europe, the concept of "free-range" chickens and hogs has captured the imagination of the buying public. For a highly urbanized population, consuming eggs and meats raised in wholesome, unrestricted natural settings has great appeal. Left to roam freely, chickens and hogs in the open country eat nutrient-rich bugs, including maggots, mice, carrion, and plants. In addition to eating naturally occurring food items, free-range chickens and hogs are thought to be less stressed and, thus, emotionally healthier. Free-range hogs have been found to have higher levels of omega-3 fatty acids and vitamin E in their meat than hogs that were raised in controlled environments.[10] For an unknown number of consumers, the thought that the animals they eat had a happier life has a good bit of appeal. While for some members of the buying public, consuming free-range animals is a humane practice, for others, the idea borders on animist spiritualism. The practice also appeals to farmers because it means that they will have to buy fewer bags of feed. The practice of raising and consuming free-range animals suggests that we are living closer to nature and will hence produce a smaller carbon footprint. In turn, a smaller carbon footprint, according to the prevailing discourse, carries with it a geomisthosis benefit: it will help to cool a heating planet. As is shown in the next section, it could also exacerbate the effect of declining temperatures if the earth experiences a return of a little or, indeed, a great ice age.

THE IMPRECISION OF PREDICTING GLOBAL WARMING AND CLIMATE CHANGE

In previous chapters, we spent a good bit of time exploring the accuracy and efficacy of making predictions to warn people of impending hurricanes, tornadoes, and earthquakes. Because of the plethora of factors that affect heat and energy flows, the ability to predict climate change is even more difficult. Unfortunately, the debate surrounding human-caused global warming not only politicized that specific subject, but it has also diminished a sense of urgency for other environmental issues that have clear and immediate impacts on biological communities. Some of the geokolasis issues that still plague the developing world are toxic gases, acid rain, ground and surface water protection, waste disposal, and erosion. Most of those issues were discussed in earlier chapters. Voices articulating concerns about these topics are hard to hear over the roar produced by participants in the global warming debate.

A major problem associated with altering the debate is the topic's relationship to overarching economic worldviews. Karl Marx argued that religious institutions and belief systems like climatic geokolasis are supportive of larger economic interests.[11] Proponents for or against the geokolasis aspects of global warming tend to also be either anticapitalist vis-à-vis Aldo Leopold or procapitalist. As with any belief system, its doctrines serve as filters through which information must travel. In the case of climatic geokolasis related to economic activities, it seems that proponents on both sides of the debate have a difficult time seeing that the remedies for other health-threatening air pollution and water use issues would have the desired impact on any possible human factor in climate change scenarios. One can also look at the debate as one that centers on the balance of power between humans and the environment or, to frame it in terms of geographical paradigms, environmental determinism versus possiblism—both have clear ways of perceiving real and imagined climatic geokolasis scenarios. As is shown in this and the last chapters, we have no choice but to embrace human ecology because we are facing a future that offers limited natural resources, including ideal and safe living spaces, so we must adapt. Will adaptation be based on real or imagined issues stemming from false premises that influence public discourse?

Arguably, erosion and groundwater contamination are just not as interesting to the general public and academics seeking grant funding from agencies and foundations as wholesale death and destruction resulting from melting glaciers, rising sea levels, and drowning cities. However, when predictions of these cataclysmic events fail to materialize, perceptions are affected and members of the public lose interest in that and, sadly, other environmental issues. Consider the implications of the warning Oliver S. Owen wrote in 1980, which, by the way, was at the end of a forty-year cooling trend:

> It is estimated that by the year 2000 the atmospheric carbon dioxide will increase another 25 percent, to almost 400 ppm. Scientists are not sure what effect this will have on human survival. Some authorities believe, however, that it might cause marked climatic changes, some of which might be deleterious.... With greatly increased quantities of heat-trapping carbon dioxide in the atmosphere, the earth's average temperature may rise sufficiently to melt the polar ice caps. The resultant increase in ocean levels would cause the flooding of populous coastal cities such as New York, Boston, Washington, New Orleans, and Los Angeles.[12]

The year 2000 came and passed without rampant flooding, and by the end of the first decade of the twenty-first century, the public in the Northeast was presented with weather conditions suggesting that such was not going to be the case in the foreseeable future. Nevertheless, other writers since 2000 have continued to focus on coastal flooding with implications for adaptation.[13] At the close of the century's first decade, the public in the Northeast observed a starkly different environmental reality. On February 27, 2010, Manhattan's Central Park received 36.9 inches of snow, breaking a 114 year old record.[14] The next winter produced similar weather events, including the blizzard of Christmas weekend in which thousands of flights were cancelled and at least a dozen people died in weather-related traffic accidents.[15] Under adverse weather conditions, authorities in New York City were forced to shut down the city's three major airports.

The Christmas storm that blanketed the Northeast had moved up the coast producing winds that reached fifty-nine mph. By Monday evening, snow had reached knee-deep height throughout the region, and drifts of up to five feet were common. In a time in which the scientific community and public were concerned about the melting of polar ice caps and coastal flooding, thousands of people were stranded in airports. With snow piling up faster than city workers could remove the white stuff, taxis and airport shuttles were unable to ferry people to hotels or their homes. Makeshift sleeping arrangements were set up in terminals. Due to the sheer volume of people stuck in remotely located airports, food was in short supply there. According to Lance Jay Brown, a traveler stranded at John F. Kennedy International Airport, "Here [in Terminal 8] there are maybe 200 folding cots for 1,000 people. I paid $50 for three hot chocolates, a couple of candy bars and two sandwiches, and I was happy to get a sandwich. There are dozens of people twisted out of shape with frustration."[16] Stranded passengers who tried to reschedule their flights experienced even more frustration. "One caller seeking to reschedule a flight on U.S. Airways was told by an automated phone message: 'your wait time is now 170 minutes.'"[17] In addition to transportation businesses (taxis, shuttles, and airlines), the United Nations canceled all events at its New York City headquarters.

A fifth winter storm descended on Manhattan and the Northeast on the night of Wednesday, January 26, 2011, and by late in the morning on the twenty-seventh, nineteen inches of snow laid on Central Park's footpaths and benches. Newark International and John F. Kennedy International airports were forced to close for ten hours, and 2,142 flights were cancelled at LaGuardia Airport. Amtrak shut down service between New York City and Boston.

In a familiar pattern, the storm had traveled up the coast, dropping snow from Virginia to Maine. In Washington, DC, commutes that normally take forty-five minutes became eleven-hour nightmares. Hundreds of drivers apparently became frustrated by the snarled traffic and abandoned their cars on major arteries in and around the nation's capital. To the delight of fiscal conservatives, government offices opened two hours later than normal. Conditions were no better in Philadelphia, where three hundred flights were cancelled and fifteen thousand households were without electricity. In fact, some 640,000 homes across the mid-Atlantic and northeast regions of the country were without electricity.[18] The storm broke January snowfall records in Connecticut. Bill Simpson of the National Weather Service said that, as of Thursday morning, the airport at Windsor Locks had received 54.9 inches of accumulation for the month. That amount of snow not only broke the previous monthly record of 45.3 inches set back in December 1945, it obliterated it.[19]

Those significant snowfalls created obvious problems for commuters and people who depended on electricity to heat their homes. Managing traffic arteries and repairing downed power lines were expensive undertakings for taxpayers and utilities. Exorbitant costs associated with coping with the storm of January 27, 2011, forced New York City to spend the last dime of its $38 million annual budget for snow removal and winter maintenance.[20]

Southern states were not spared either. From January tenth through the thirteenth, 2011, snow and ice accumulated on highways, tree limbs, and lawns from Louisiana through the Carolinas and on up the coast, where as much as two feet of the white stuff fell once again on parts of the New England states. In some neighborhoods of Charlotte, North Carolina, five inches accumulated on Bermuda grass lawns. In Atlanta, Georgia, the weather was not exactly "peachy" either. Four to seven inches of snow forced Atlanta's Hartsfield-Jackson International Airport, the busiest in the world, to cancel most of its flights. In South Carolina, snowflakes looked odd falling on palmettos, but they were dangerous, accumulating on highways and back roads. On Monday, January 10, state troopers worked two thousand traffic accidents. In total across the Deep South, the snow storms of January 10–13 caused at least nine people to lose their lives in traffic accidents.[21]

Other sections of the contiguous states were hit by successive winter storm events. The central plains states were certainly among the hardest hit. On February 8, 2011, the second blizzard in one week caused the cancelation of flights in and out of Dallas, Tulsa, Oklahoma, with more than one month of winter left on the calendar, broke its winter accumulation record originally set back in the winter of 1923–1924. "In Tulsa, where mail, bus, and trash pickup service was only recently restored, 5 inches of new snow gave it 25.9 inches for the season, breaking its old seasonal record of 25.6 inches. . . . Last week's record 14-inch snowfall kept students out of school for at least six days and made many roads in the state's second-largest city impassable."[22]

While not trying to minimize the ability of Americans to make abstractions about weather and climate, the concrete evidence of cold weather,

Image 10.1. A snow-covered Tennessee interstate. *Image courtesy of the author.*

Image 10.2. Edisto Beach, South Carolina, on December 29, 2010. Looks are deceiving; it was 38°F. *Image courtesy of the author.*

Image 10.3. Georgia conifers covered in snow in January 2011.
Photo courtesy of Margie Langley.

along with the inconvenience that such conditions create for those who have to travel, reduces a sense of urgency to act quickly to cool a planet made red-hot by human actions. In effect, recent cold weather in the developed Northern Hemisphere has reduced the dreaded geokolasis aspect of global warming. The winters of 2010 and 2011 also illustrate how difficult it is to make reliable climate predictions. Indeed, a Gallup poll taken in late March 2011 shows that the public's concern for environmental issues has declined since 2000. Of the nine environmental issues about which Gallup queried people, water-related risks and issues were the most worrisome. Global warming and open spaces were the least of the public's concerns.[23]

Among the challenges in making predictions about climate change is that the correlation between atmospheric carbon dioxide and temperature gradients is not well established.[24] There are multiple factors related to human activity besides burning fossil fuels that theoretically can increase air temperatures, including other greenhouse trace gases like methane, nitrous oxide, chlorofluorocarbon, tropospheric ozone, and humidity. On the other hand, there are other human-related factors that can have a cooling effect on climate, which include airborne particulates or aerosols such as sulfate particles, organic compounds, and soot. These aerosols scatter or reflect shortwave radiation from the sun back into space. Chlorofluorocarbons can also have a cooling effect because, in the stratosphere, they destroy ozone.

Natural processes also influence atmospheric temperatures, including greenhouse gases. Volcanic eruptions spew gases and aerosols that can either heat or cool the atmosphere. As was described in chapter 7, 1816 is referred to as "the year without a summer." Cold conditions that year were caused by the eruption of Mount Tambora on the island of Sumbawa in Indonesia in 1815. Aerosols produced by that eruption reflected enough sunlight into the stratosphere to cause a drop in global temperatures of between 0.4°C to 0.7°C (0.7°F–1.3°F).[25] A similar eruption occurred in the spring of 1991 at Mount Pinatubo on the island of Luzon in the Philippines. That eruption inserted approximately twenty million tons of sulfuric acid aerosols into the stratosphere. Over the next two years, global temperatures cooled by about 0.3°C (0.5°F).[26] As we will see shortly, huge quantities of greenhouse gases are also emitted in eruptions.

Climate can be affected by other natural processes as well. Periodic changes in the earth's tilt can affect the angle of sunlight penetrating the atmosphere. More direct sunlight has a heating effect on climate, and, conversely, less direct sunlight has a cooling effect. Also, solar activity varies from time to time. Additionally, cloud cover over tropical waters can reduce the amount of energy that is dispersed into zones of energy deficit that extend toward the poles from about 40° north and south. Because it is so difficult to accurately weigh each of these man-made and natural factors, some scientists believe that recent global warming is part of a natural process of climate oscillations.[27]

Even looking back over time yields researchers little more insight into how to establish a temporally based estimate for climatic oscillations, although as we will discuss shortly, at least two scientists believe there are thirteen hundred intervals between mini ice ages. One important fact is revealed by looking back through the pages of history and through the geologic record: climate change happens and it will continue into the future. Ironically, higher levels of atmospheric water and carbon dioxide may actually keep the earth's climate from slipping back into a Pleistocene-like epoch. Not only would a recurrence of an ice age spell certain doom for European cities like Moscow, Berlin, Geneva, Oslo, Manchester, and Birmingham, among others, North American cities like Montreal, Toronto, Minneapolis, Lansing, Buffalo, Lincoln, and Albany would suffer the same destructive fate. In addition, most of the fertile breadbasket of North America would be lost, and the growing season in the southern breadbasket of the continent would be shortened. Given current technology, our ability to feed billions of people would be diminished.

THE HISTORICITY OF CLIMATE CHANGE

As suggested above, higher levels of greenhouse gases like humidity and carbon dioxide could spare humanity from an unpleasant, frigid future. To fully appreciate that thought, it is necessary to partially challenge the orthodox view of dynamic equilibrium with respect to the carbon and hydrologic cycles. The fact that water and carbon are recycled is beyond dispute, but the notion that nature, absent human interference, balances the amount of carbon added to the atmosphere and the amount that is removed is not supported by tangible evidence. Indeed, natural systems, if left alone, remove vast quantities of water and carbon from circulation, which, in the long run, would have a cooling effect on climate and may help explain why there have been eight distinct "ice ages" in the past 720,000 years.[28]

In the case of water, sandstone aquifers such as the Ogallala Aquifer that lies beneath the fertile High Plains of the United States once held trillions of gallons of water that were removed from the hydrologic cycle. However, mining water to irrigate range- and croplands as well as supplying municipalities with the resource is returning significant amounts of water to the hydrologic cycle. Also, glaciers retain water that has not been in gaseous form for millennia. Carbon, on the other hand, begins its journey in supporting life as gaseous carbon dioxide (CO_2). When CO_2 enters a plant through its stomata or breathing pores, sunlight initiates a process called photolysis that splits water molecules into oxygen and hydrogen whereupon carbon is fixed with hydrogen to form sugars. Oxygen is an important byproduct that escapes through the plant's breathing pores, and carbon produced in plants is passed through the food chain.

Like water, nature continually removes carbon from its cycle. In addition to trapping carbon in coral reefs, billions of tiny globigerina live out their short lives in the world's oceans. Their remains fall from the surface of the seas like an organic rain, and their calcium-carbon-based skeletons accumulate on the ocean floor. Ralph Buchsbaum estimated that about 30 percent of the ocean floor (some forty million square miles) is covered in grey globigerina ooze.[29] In deeper parts of the ocean, the ooze eventually forms chalk, and in shallow coastal areas, the ooze lithifies into limestone. Clearly the formation of coral reefs and the buildup of chalk and limestone layers provide evidence that carbon is not always recycled and used in the circle of life. Further evidence of this process is shown in the formation of coal and crude oil deposits.

The Carboniferous age was a phase of the Paleozoic era. It lasted from about 290 to 354 million years ago. During that time period, the earth's climate was warmer and wetter. Lush forests, including giant tree ferns, grew in much of the eastern portion of what is now North America, especially in Ohio, Pennsylvania, Virginia, West Virginia, Kentucky, Indiana, and Illinois, as well as across central and southern England and central Europe. That was also a time of tectonic mountain building caused by continental collisions. In North America, the period is divided into the Lower and Upper Carboniferous eras. Whereas the Upper Carboniferous is identified by mostly coal deposits, the Lower Carboniferous is associated with limestone, a sedimentary rock composed of calcium carbonates deposited on ancient seafloors by the remains of crustaceans and crinoids (lime-encrusted algae). The Upper Carboniferous age had a tropical climate that supported vegetation similar to the plants found in the Orinoco Delta of eastern Venezuela.[30]

The effects of continental or plate collisions must have been dramatic because sediments quickly covered large quantities of carbon-based plant matter. In doing so, huge quantities of carbon were removed from the circle of life. In more scientific terms, a significant portion of the elemental cycle

was removed from the biosphere. Those important events in the earth's natural history hint at a process that continued through the ages and likely led
to the Pleistocene, the last great ice age. By removing huge quantities of
carbon dioxide and humidity from the biosphere, nature cooled itself and
killed off countless species of plants and animals that could not adapt.

The Holocene, the relatively warm interglacial period in which we find
ourselves, is only about ten thousand years old. The Holocene is also known
as the Age of Man because all of recorded history has occurred during this
time period and because human population climbed from fewer than five
million to more than seven billion. If the Holocene's proximity to advances
of continental glaciers is correct, the earth's climate will return to a much
colder state. However, there is no way of knowing when or if another ice age
will occur. Nevertheless, the return of an ice age would be devastating to
humanity. With a drastic reduction in the earth's culturally modified carrying capacity, social order would collapse and the Malthusian Trap would
be activated with unimaginable consequences. Aspects of this unsavory
topic are examined further in the final chapter, but for now, let us return to
a temporal delineation of climate change and how releasing carbon dioxide
and humidity back into the atmosphere may not be such a bad thing. This
is important to consider because most of the contemporary thinking on climate change has focused on warming as a bad thing.[31] Such thinking is not
likely to change any time soon because the year 2009 ended the warmest
decade on record.[32]

Since 1880, the year in which reliable thermometers were put to use at
various places around the planet, global temperatures have risen by 0.8°C
(1.5°F).[33] However, between 1940 and 1980 temperatures declined almost
to their 1880 level, and as we learned previously, winter storms in February
and December of 2010, as well as January 2011, dumped record-breaking
snow falls on Connecticut and Manhattan and created havoc and economic
disruption throughout the Northeast.[34] The blizzard of Christmas
weekend in Philadelphia, Pennsylvania, was so severe that an NFL football
game scheduled for December 26, 2010 was cancelled. "'Mother Nature is
trying to wreak havoc on the NFL schedule—and it involves the Minnesota
Vikings® again,' a post on NFL.com read."[35] The massive snowfall and asso-

ciated cold weather was not limited to the Northeast. Earlier in the month, on the night of December 12–13, some eighteen to twenty inches fell on Minneapolis, Minnesota, a city made partially famous for its innovative Mall of America and the "indoor effect" it has created in the downtown portion of the city. Another of the city's "weatherproofing" innovations was a dome, handily named the Metrodome, built for its Vikings football team. The fiberglass, Teflon®-coated roof collapsed that night under the weight of eighteen inches of snow.[36] Fortunately, it was not during game time. On that same weekend, snow flurries fell on NFL players in Jacksonville, Florida, as the game time temperature remained 25°F below the average high of 60.3°F.[37] Admittedly, the snowy events that occurred in the successive winters in 2010 and 2011 do not indicate that an ice age is returning, but they do suggest that weather patterns are not linear. Change does happen. A longer look at the earth's past supports that contention.

As was shown in the chapter 4 discussion on geotheology from the cradle of civilization, imagery of weather and climate in the Bible suggests that conditions in the Near East were cooler during the time in which the book of Job was written (ca. 950 BCE). Nonbiblical evidence for cooler climate conditions during that time period is suggested in the research conducted by geologist Lloyd D. Keigwin who studied salinity and the flux of terrigenous material (weathered rock from continental sources) in the Saragossa Sea, which, despite being situated west of the Mediterranean Basin in the middle of the North Atlantic, nonetheless shares a similar latitudinal location as the northern Near East. Keigwin found that temperatures prior to medieval times were 1.0°C (1.8°F) cooler than they were in 1996.[38] In a similar manner, nonbiblical support for icy conditions in the Near East during that time period is provided by paleoclimatologists Charles A. Perry and Kenneth J. Hsu who creatively tied together geophysical, archeological, and historical evidence to support a solar-output model for climate change. They identified thirteen-hundred-year cycles of climatic cooling and heating during the Holocene. That cycle for little ice ages was only skipped over once, around 7500 BCE.[39] The person who wrote the book of Job was clearly witnessing a cold climate or a little ice age. Perry and Hsu describe in good detail a little ice age that lasted from 1250 BCE to 750 BCE, a time frame that most likely coincided with the writing of the book of Job. Assigning a date for the book of Job, however, is difficult, but theologians and Bible editors Earl D. Radmacher, Ronald B. Allen, and H. Wayne House, argue that the mention of iron tools (19:24, 20:24, and 40:18) suggests a date after 1200 BCE. More specifically in Job 7:17, they point out that the author seems to allude to a passage in Psalms (8:4), so it is quite possible that Job was written during the time of Solomon (ca. 1017–931 BCE).[40]

Perry and Hsu go on to argue that the little ice age that befell the planet between 60 and 600 CE was an environmental force behind what they call the "migration of nations."[41] In particular, they point out that it was during this time that ancient Germanic peoples such as the Vandals and Goths, and Asiatic folk like the Huns besieged the Roman Empire. In Asia, the Chinese Empire was likewise overwhelmed by Asiatic peoples. On the other hand, this little ice age helped Central America's Mayan civilization flourish on the Yucatan Peninsula.[42] Whereas Germanic and Asiatic tribes migrated to warmer environs where they ran into conflict with existing civilizations, the cooler and drier weather in the Yucatan drove malaria-carrying mosquitoes farther south. The tropical climate gave way to a more temperate weather pattern that was better suited to cultivation and city building. However, when the Medieval Warm period (also known as the Medieval Optimum) arrived, so did the mosquitoes carrying the scourge of malaria.

The Medieval Warm period, which ran from circa 800 to 1250, is well documented in Scandinavian migration history, but in scientific literature, its existence has been disputed by some scholars while others dismissed it as a minor regional phenomenon.[43] However, as scientists from the Southern Hemisphere have joined in the quest to challenge the historicity of the Medieval Warm period, it is clear now that it occurred in both the Northern and Southern Hemispheres.[44]

The Little Ice Age that ran from 1280 to 1860 was perhaps the coldest period on earth since the retreat of the last continental glacier some ten thousand years ago. This was also a time of the bubonic plague and the resulting Black Death that swept across Europe and arrived in the British Isles in 1348. Also around this time, the river Thames routinely froze and was sufficiently safe enough for people to skate on it. In Europe, alpine glaciers extended well into valleys inhabited by people. In the Americas, the Cherokee, who spoke an Iroquoian language, migrated into southern Appalachia from the Saint Lawrence River area of upstate New York and southeastern Canada. The Little Ice Age was also arguably responsible for the migration of the Shawnee into southern Ohio and Kentucky. The Shawnee spoke an Algonquin language and were cultural relatives of the Algonquin tribe of Quebec and Ontario. While there is no written record of the reasons for their relocation, the Cherokee obviously left behind their Iroquois relatives in the Northeast to resettle in what are today North Carolina, Tennessee, and Georgia. Given the fact that the Little Ice Age had reduced the regional carrying capacity in eastern Canada and upstate New York, it seems logical to conclude that their migration was pursued for the sake of survival.

An argument can also be made that the Little Ice Age played a role in pushing Europeans to migrate across treacherous water bodies. The settlement of the Americas by Europeans obviously occurred during the Little Ice Age. The colder climate impacted even cross-channel migrations between Britain and Ireland.[45]

Temperatures during the Little Ice Age would have varied in accordance with latitude, elevation, and proximity to large water bodies. Locations on the windward side of a continent or large island like Greenland, for instance, would have seen less of a change than places located on the leeward side of the island or continent. Tropical places with low elevations like the Yucatan would have enjoyed a temperate climate that may well have resembled Mount Sandel (7000 BCE) or Skara Brae (3000 BCE) during ancient and long-forgotten warm periods.[46] On balance, however, the average temperature in the North Atlantic, according to Lloyd D. Keigwin, was 1.0°C (1.8°F) lower than it was in the 1990s.[47] Again, this needs to be seen in light of the fact that the range of temperatures is always lower in marine climate zones than it is in continental locations like the Great Plains or Siberia.

Studies conducted in South America's Patagonia show that there was a cold, moist climate in the region from 1270 to 1660.[48] Farther to the west, in the South Pacific, geochemist Erica J. Hendy and a host of her colleagues conducted a study that delineated a 420-year history of strontium/calcium, uranium/calcium, and oxygen isotope ratios measured in eight coral core samples taken from the Great Barrier Reef located off of the northeast coast of Australia in the Coral Sea. The results of their extensive study showed that salinity was higher in the mid 1500s through the middle of the nineteenth century than it was in the twentieth century. Hendy and her team concluded that the sea "freshened" because warmer temperatures toward the poles after 1870 reduced the velocity of winds over the tropics and, hence, lowered the amount of moisture evaporated from the waters off of the coast of Australia. As a result of lower evaporation rates, they reasoned, more water was available to dilute sea salt in the tropical latitudes of the South Pacific.[49]

Between 1880 and 1940, global temperatures rose slightly above the world-wide average temperature of 59°F. A gigantic eruption on the East Indian island of Krakatoa in 1883 spewed particulate matter into the atmosphere from May twentieth until August. Sunlight striking airborne matter from Krakatoa was reflected back into space, and, as a result, global temperatures were markedly lower throughout the next year. Its effects on weather were strong in the Northern Hemisphere. Bostonians were able to throw snowballs in June. Nevertheless, the phase of warming reached its maximum effect during the late 1930s, at which time overgrazing and deep-plow cultivation in the western Great Plains contributed to the Dust Bowl.

From around 1940 through the late 1970s, global temperatures declined to just below the global average.[50] Global temperatures dropped to a near-record low in the wake of the eruption of Mali's Mount Agung in 1963.[51] Almost thirty years later, on July 16, 1990, the island of Luzon in the Philippines was rocked by an earthquake that registered 7.8 on the Richter scale. The epicenter was sixty miles northeast of Mount Pinatubo. The earthquake was a precursor to a buildup of volcanic gases that caused a violent eruption of Pinatubo on June 15, 1991. The gaseous explosion expelled one cubic mile of sunlight-deflecting material into the atmosphere. Throughout the remainder of 1991 and on through 1993, global temperatures dropped by 0.5°C (about 1.0°F).[52] Thanks in large measure to the predictions made by the Philippine Institute of Volcanology and Seismology and the US Geological Survey, at least five thousand lives were saved and $250 million in property damage was averted.[53] While the US Geological Survey and writers like Oliver S. Owen give ample attention to the cooling effects of particulate matter spewed from volcanic eruptions, they seldom note the possible climatic connection to the huge volume of greenhouse gases that are belched out into the atmosphere. The most abundant gases usually released in an eruption are H_2O followed by CO_2, followed by sulfur dioxide (SO_2).[54] Particulate matter settles out of the atmosphere long before nature can down-cycle sudden bursts of CO_2 from eruptions like Pinatubo and Mount Saint Helens in 1980. Given the fact that the prevailing weather pattern from 1980 to 2009 was one of increasing temperatures that reached a peak in the first decade of the twenty-first century, the role in global warming scenarios played by greenhouse gases emitted by volcanic eruptions deserves further study.

December 2009 ended the hottest decade on record, but climate could cool down again. Indeed, as was shown in the section of this chapter on the imprecision of predicting climate change, weather conditions in the winter months of 2010 and 2011 took a nosedive. As mentioned above, cities like New York and Boston in the Northeast were not inundated by rising sea levels caused by melting ice caps and glaciers, but they were hard hit by successive snow storms and blizzards. The winter rampage was not limited to the Northeast and the contiguous states. In the United Kingdom, 2010 witnessed the coldest December on record. However, record keeping on a national scale in the United Kingdom only goes back one hundred years, to 1910.[55] Nonetheless, December 2010 was not just a little colder than usual. The long-term average for the month was 4.2°C (39.56°F), but it dropped to −1.0 (30.2°F) in 2010. Elsewhere in the United Kingdom, December temperatures were at an all-time low, such as in Northern Ireland, and they were at their lowest in Scotland since 1947.[56]

While these events do not necessarily suggest that a mini ice age or a great ice age is looming, they should remind us that predicting climate change is fraught with a host of problems. This should also remind us that the return of a major ice age could happen, but, as research by Perry and Hsu suggests, it is more probable that a little ice age will reappear sometime

Image 10.4. Perth, Scotland, December 2010. *Photo courtesy of Colin Williamson.*

Image 10.5. Icy Northern Ireland, December 2010. *Photo courtesy of Colin Williamson.*

during the next millennium. However, given that we live in the Holocene, which is an interglacial period, a great ice age is due to return. It is a paradox to think that human activities, so lamented by those who fear climatic geokolasis, may actually keep midlatitude expanses of the earth habitable for centuries to come.

SUMMARY

The historical nature of climate change and the tendency for climate over the past one million years or so to move back and forth between ice ages and interglacial warm periods should get more attention than it receives in academic and popular literatures. Global warming has dominated the discourse, and the geokolasis fears associated with it have caused some proponents of human-caused climate change to call for radical changes in economic activities and even for control of individual and collective behaviors. It follows that many members of the public and the academy are willing to sacrifice personal liberties to protect the planet from our exploitive activities. As suggested by cultural geographer Mike Hulme, proposals for governance on a global scale are brash, utopian, and not very practical.[57] Indeed, they are little more than examples of the historic tendency of people to offer repentance for sinful behaviors that invited the wrath of nature to befall wayward humanity. Whereas our ancient and recent ancestors saw these scenarios as geotheokolasis events, more secular people today have expunged God, but they still maintain geokolasis beliefs that do not lend themselves to finding rational solutions for a host of other environmental problems that affect air and water quality.

As we look ahead, our focus should include contingencies for the return of an ice age. Climate change presents many challenges, but, as will be shown in the final two chapters, all contingencies for coping with these and other formidable forces of nature need to be made with respect to the widest possible participation of individuals, groups, and societies. Our survival as a species has always depended on adaptations made by micro- and macrosocial organizations. Just as our federal system allows for fifty distinct experiments in democracy to try out novel ideas, the world's 194 countries and their communities offer the best chance for identifying ways to adapt to the challenges of climate change. However, not all countries share the same worldview or an equal understanding of the state of our ecumene. Perhaps the greatest challenge we face is in nurturing an international concern for cooperation on mutually pressing environmental issues.

Chapter 11

THE EAST'S HUMAN TSUNAMI

INTRODUCTION

Despite the fifty thousand or so years that humans have been moving away from the Near East, the tide of humanity from its original homeland shows no sign of abatement. However, over the millennia, flows of migrants have changed in regard to the types of settlements that attract them as well as their geotheological imagings. From hunting and gathering pantheists to successive waves of skilled agriculturalists adhering to monotheism and polytheism of various stripes, the spring of humanity is not dry. As members of smaller communities with limited knowledge of people not like themselves, personal identities and broader affiliations stemming from tribes, clans, and nations have evolved. These social and personal identities are adopted to help make sense out of incredibly complex and diverse natural and cultural environments in which maturing people find themselves.

More recently, even the economic character of migrants has changed yet again from that of a small-settlement agriculturalist to an urban dweller with inclinations to develop knowledge and skills in line with evolving economic niches. As a result, most coastal cities in the Northern Hemisphere, called "the land hemisphere," are teeming with people who need food, clothing, and shelter produced or fabricated by members of rural communities. With four of the five most densely populated regions in the world located on the Eurasian landmass, continued population growth in the ancient ecumene will certainly place billions of people in harm's way. While western Europe is, for the time being at least, environmentally stable, its

243

northerly location and the possibility of a return of an ice age makes the quality of life there over the next century less certain. Given that 60 percent of the world's population lives in Asia either near the Bay of Bengal and the Indian Ocean or along the Pacific Rim, there are a host of natural disasters awaiting future generations of southern and Southeast Asian peoples.

In the meantime, increasingly dense populations in the ancient ecumene will have to rely on more sophisticated and technology-dependent economic systems to survive. To meet their needs, respective societies must expand segments of the earth's carrying capacity through changes in economic structures. In places that were once natural utopias, large cities have grown and people living within them are increasingly faced with survival choices. Unlike their ancient ancestors who could move on down the coast without making many changes in their way of life, modern folk are faced with a more challenging ordeal if they decide to move away from familiar environs. For instance, they often find that they will have to relocate to societies that speak different dialects or even tongues that are not part of their native language family. They may also find that their religious worldview is not widely shared by many in their adopted country.

As the prevailing global migration pattern is dominated by relatively poor people from highly fertile countries moving to economically diverse societies, their attitudes about traditional families, gender roles, and the number of children they should have are often at odds with the larger, and in many cases mainly secular, adopted society. One common adaptation that families have made in the more developed world is to have fewer or no children. The deliberate reduction of natural increase rates (crude birth rates minus crude death rates) in these places suggest that the human ecology paradigm has merit and that reproductive choice is a reflection of larger cultural adaptations to environmental limitations. This is an alien and quite possibly heretical notion in many of the world's poor countries.

Nevertheless, it is the more complex and economically diverse settlements that draw migrants from cultures that cannot support them. Families not exposed to the social and cultural innovations brought on by the Industrial Revolution of the eighteenth and nineteenth centuries depend on the labor of both parents and children to survive, but evolving niches in places that have experienced industrialization no longer regard children as assets. In more developed countries like Sweden, Italy, and Japan, women's roles are no longer seen as well served at home where they traditionally acted as caregivers and nurturers of the young. They are now regarded as integral workers in highly diverse and specialized economies. With changing gender roles, families in more developed countries are producing fewer and fewer children.

Thanks in large measure to these changes, a relatively small number of countries have achieved the level of economic development found in Italy and Japan. While Italy has a population density of 512 people per square mile, Japan is the home of 879 people per square mile. While each country has its own unique religious and cultural backgrounds, they share a number of similar geographic traits.[1] One occupies the majority of a peninsula in southern Europe, the other is spread out over a series of islands located off the coast of east Asia. Each country sits on the edge of a continent, and each has previously sent forth waves of migrants, which no longer seems to be their first option for coping with surplus populations. Arguably due to economic considerations, restricting child birth has become the most efficient way to live in their urban environments. Declining natural increase rates in more developed countries like Japan and Italy is arguably a response to stresses on regional carrying capacities and is hence a manifestation of a way those countries cope with the limiting forces of nature. However, their lower and even negative natural increase rates create a dependence on immigrant labor, and the Near East, as it has for fifty thousand years, continues to supply distant lands with humans.

DEMOGRAPHIC ADAPTATIONS IN MID-ATLANTIC COASTAL AREAS

Without immigration, the demographic future does not bode well for Europe. Indeed, economic conditions in the form of austerity measures enacted by governments in Greece, Spain, and Ireland, as well as the United Kingdom, suggest that tax revenues are drying up along with investments in domestic production. As we will see later in this section, Asian countries are becoming industrial powerhouses. Nevertheless, for the next few decades there is no reason to believe that Europe's current demographic trends are likely to change. Let us consider the changes that are occurring in the European population. There were 186 million people living in western Europe in 2005, but based on population trends that number will drop to 183 million by 2050.[2] Southern Europe will also see a drop in population. The projected number of 141 million for 2050 reflects a 6.6 percent decline. The situation is even worse in eastern Europe. The present population of 297 million will fall to 232 million by the middle of the twenty-first century. Meanwhile, in the Near East and other countries located across southern Asia, the opposite trend is occurring. For instance, in southwestern Asia where we find countries like Kuwait, Turkey, and Saudi Arabia, the current population of 214 million will climb to 400 million. In North Africa, the 2005 population will climb from 193.8 million to 323.6 million by midcentury. Later in this chapter, we will take a closer look at growth among all countries located along the southern shores of the Persian Gulf, Bay of Bengal, Indian Ocean, and Pacific Ocean. Suffice it to say, these trends are nearly universal across the Near East and southern and southeastern Asia.

What these figures do not tell us is that changes in population represent growth among some of the world's most economically disadvantaged people while more affluent and better-educated countries are aging and ultimately dying. Urbanism, which characterizes the social environment of the latter, produces a demographic profile that is like a cancer. It eats away at the most developed societies, creating a need to recruit foreign workers to fill less desirable employment niches. Like millions of people before them, immigrants from the Near East are simply filling niches in new places, but unlike those who migrated during the Mesolithic and early Neolithic ages, diversified economic structures require a good deal of cooperation among charter residents and new arrivals to efficaciously fit into the local carrying capacity. Otherwise, social upheaval and political conflict will cause a host of problems.

It is important to recognize that international immigrants seek survival opportunities in certain economic niches in other countries, so they are resistant to change their culture and worldview. Social welfare programs in more developed societies represent an attractive pulling force for people living in comparatively poorer countries. However, if a restricted number of economic niches including social welfare programs attract immigrants, then any future reduction in employment in those sectors or cut backs in welfare programs will result in social conflict between immigrant groups and charter residents. Such was the case in 2010 as countries like France, Greece, Spain, and the United Kingdom, faced with excessive expenses to care for increasingly dependent populations, enacted austerity measures to ensure their governments' solvency.[3] While cultural diversity may be a societal asset when accompanied by full employment, it is a bane to societal solidarity when times are tough. The presence of ethnic enclaves and a process of in-grouping and out-grouping based on forces of disunity that are difficult to identify and manage will exacerbate tensions. The Malthusian scenario of social conflict, including wars, famines, and epidemic diseases, will be a distinct possibility for much of the populations living along the Pacific Rim and in southern Asia.

Few in the West realize that these social manifestations are more a result of shifts in global investments away from Western countries and toward the human-resource-rich countries of the East than they are because of draconian motives of politicians who propose austerity plans to ensure their countries' survival. Ethnic diversity, which already characterizes life in economically vibrant countries like Malaysia and Indonesia, will increase among Asian countries as their carrying capacities expand through economic diversification and technological innovations. Such was the case in the United States after the Civil War and in Europe after World War II.

As Europe rebounded from the devastation leveled on it during WWI and WWII, economic expansion coupled with a general decline in the natural increase rate for a number of countries combined to create a labor shortage. By the 1960s, a number of western European countries openly recruited guest workers from Spain, Portugal, and Italy, but, as that supply of relatively cheap labor dried up and labor shortages continued to vex their economies, European companies began to recruit workers from the Muslim areas of Yugoslavia, Turkey, and North Africa.[4] By the late 1970s, western Europe contained eleven million foreign-born residents, representing about 5 percent of the population. The vast majority of the Muslim immigrants settled into the older sections of existing urban centers like Brussels, Paris, and Berlin. By the late 1980s, Muslim populations in those cities were 16 percent, 12 percent, and 18 percent, respectively.[5] More recently in France, population data published in the mid-1990s showed that 90 percent of the population was at least nominally Roman Catholic and the Muslim population was less than 2 percent.[6] In 2006, the Roman Catholic population had dropped to 83 percent while 10 percent of the French population was identified as Muslim.[7] In 2001, some 27 percent of London's population was born outside of the United Kingdom, and nearly one-third of Londoners were identified as non-white.[8] In London since the 1970s, there are sizable clustered settlements of Muslims, particularly in Newham and Tower Hamlets.[9]

As most of the surplus wealth in the developing countries of the Near East and of southern Asia will be consumed to take care of the young and elderly or will contribute to increasing the wealth of the elites, their economies will likely experience little, if any, expansion. These countries' inability to employ more people will result in a steady stream of immigrants, first to Europe and then to the Americas, as well as to east and Southeast Asia.[10] To demonstrate this point, consider that the civil unrest in Egypt in 2011 that led to the resignation of Hosni Mubarak was certainly spurred on by poverty. As we read in chapter 3, Cairo has over eighty thousand residents per square mile as compared to New York City's twenty-seven thousand people per square mile. There are too many people for too few jobs. In the absence of any major cultural transformation, which would necessarily have to radically change the role played by women in the workplace and the home, there is no end in sight to the flow of immigrants from the Near East. With fewer Europeans being born who would be able to fill employment niches to support their aging and highly dense coastal populations, there will be hardly any economic reasons to keep able-bodied workers out of their continent.

While the reasons for transoceanic migrations in a modern context are discussed later on in this chapter, it is nevertheless important at this juncture to point out that the United States is also experiencing significant settlement of Near Eastern and south Asian transplants in a number of places. The Pew Research Center estimated that in 2007 there were 2.35 million Muslims living in the United States. Of that number, about two-thirds were foreign-born immigrants. Other published estimates put the number higher at nearly five million.[11] These data along with other reports show that the rate of growth among Muslim immigrants is high. According to demographers Barry A. Kosmin and Egon Mayer at the Graduate Center of the City University of New York, who reported data on American religious affiliation generated from surveys they conducted in 1995 and again in 2001, the Muslim population more than doubled in just six years.[12] In 1995, there were approximately 527,000 Muslim Americans, but in 2001, there were 1,104,000. In other words, during each of those six years, the United States added an average of 96,166 new residents who claim adherence to Islam, which happens to approximate the population of Dearborn, Michigan, in 2006. However, since 2001, the Muslim population in the United States has grown by 207,666 people per year. That number is two times the size of Evansville, Indiana, and somewhat larger than Des Moines, Iowa. If the doubling time continues at the current rate of six years, which is a valid demographic premise, by the year 2052 there will be over 300 million Muslims living in the United States.

While many of these immigrants are relocating to Western cities that are religiously diverse and certainly more secular than their Near Eastern homelands, they are nevertheless encouraged to do so by geotheological imagings depicted in the Qur'an. Like migrants adhering to all Abrahamic religions before them, they are relocating to take advantage of the bounty that the God of Abraham has provided them in those new lands. As Muhammad wrote in the Muslim holy book, "Surely (as for) those whom the angels cause to die while they are unjust to their souls, they say: In what state were you? They shall say: Was not Allah's earth spacious, so that you should have migrated therein? So these it is whose abode is hell, and it is an evil resort."[13] Muslims are further assured that it is Allah's will for them to move to other places when need presents itself. As related in the Qur'an, "And whoever flies in Allah's way, he will find in the earth many a place of refuge and abundant resources; and whoever goes forth from his house flying to Allah and His Apostle, and then death overtakes him, his reward is indeed with Allah and Allah is Forgiving, Merciful."[14] Also, "with respect to those who fly after they are persecuted, then they struggle hard and are patient, most surely your Lord after that is Forgiving, Merciful."[15] For Muslims, passages like these can certainly lessen some of the anxiety of migrating to a foreign place.

ECOLOGICAL FORCES BEHIND INTERNATIONAL MIGRATIONS

The reasons why humans move in a modern context have received a good deal of attention from writers and, since mobility involves space, geographers, who are especially interested in spatial issues, have developed some good theories that have relevance in explaining at least contemporary migration flows. Generally people migrate because they are either pushed or pulled to and from places. As James Rubenstein points out, push and pull factors that precipitate mass migrations typically fall into one or more of the following reasons: economic, environmental, political. But I and a few other writers like Kerby Miller and Avihu Zakai would add religious beliefs to this list, although admittedly the restriction or requirement of religious behaviors in public spaces is a function of political conditions found in a given society.[16] In this instance, I am referring more to the sense among some people that their migration is ordained by divine forces. This impulse was certainly detectable in the resettlement of Palestine by European and American Jews, as well as the colonization of America by Puritan settlers from Great Britain and Ireland. Also, for some, the settlement of lands west of the Mississippi River during the era of Manifest Destiny was, as the name of the time period suggests, allowed by the sovereign will of God.

It is not surprising that in the wake of the Industrial Revolution, which caused an unprecedented flow of rural residents into urban areas in search of work, geographers and other social scientists were interested in explaining those movements and their consequences on the lives of charter residents as well as immigrants. Early giants in the field of sociology such as Auguste Comte, Émile Durkheim, Karl Marx, and the next generation of sociologists like Ferdinand Tönnies were keenly interested in the social consequences and underlying causes of movements within their respective societies. As one would expect from sociologists at the time, their interests were built on assumptions about why people had moved, which they assumed were relegated to simple and presumably obvious economic factors such as getting a job. Geographers, whose interests were limited to spatial aspects of measurable or observable phenomena such as the decline of population in one area and an increase in the numbers of people moving into another place, also developed a niche.[17] One geographer in particular established his career and legacy on what he called "laws of migration." His name was E. G. Ravenstein (1834–1913), and his laws were based on patterns he observed in Europe during the 1870s and 1880s.[18] Some of his laws have not sustained the test of time, but five have proven reliable statements for geographers in the twenty-first century. Although these statements are adequate descriptions of migration patterns, they are not truly laws. Nevertheless, here are Ravenstein's five most sustaining migration descriptors:

1. Most migrants move only a short distance.
2. Long-distance moves favor big city destinations.
3. Most migrations proceed in a step-by-step fashion.
4. Most migration flows are from rural to urban areas.
5. Migrations produce a counterflow.

We should also add that international migrations typically occur among the best-educated members of a society, which often results in a "brain drain" problem for the home country.[19] For example, among the one million Asian Indians living in America, close to three-quarters of them have a bachelor's degree. A British government study found a similar pattern among African immigrants who hold at least a high school diploma.[20] It is reasonable to expect that as the best educated and most able immigrants become established in a new ethnic enclave, even if the enclave is in another country, others less able to market themselves follow and fill lower social niches, including those created by welfare programs, in the new enclave. This process is called chain migration. These new social environments, therefore, become increasingly stratified by members of the home country. Here we should also reinsert Ravenstein's fifth law stating that migrations produce counterflows. For various reasons, some of the immigrants will remain in the host country, so counterflows are smaller than the original waves of immigrants.[21] With respect to Muslims living in other countries, there are a number of factors that make the initial flow larger than the counterflow. As Turkish economists Nil Demet Güngör and Aysit Tansel found in their study of Turkish Muslim emigrants, wage differentials, instability in the Turkish economy, work experience in Turkey, and a favorable opinion of host country lifestyles played major roles in deciding whether or not immigrants would return to the country of their birth.[22]

It is interesting to note that little attention is given to the strength of community, based on the "old country," in the host nation. If all economic factors that propel people to go in search of sustenance are equal and if the new enclave looks, smells, and feels like home, why return? Clearly government policies that help immigrants preserve their old culture and community in ethnic enclaves will increase the likelihood that they will feel at home in the host country. Multiculturalism, as a political policy, will certainly reduce the intensity of Ravenstein's law declaring a reverse flow. Contact with the old country can be maintained through social networks held together by the Internet and other communication technologies. The importation of goods from the old country, which can enrich the host country with more variety in cuisine, music, and even personal dress, also serves to make immigrants feel at home in a new land. However, just as tantalizing and appealing aspects of culture can be diffused into the host country, so can the bad. As less talented and able members of a source country join the migration chain, the likelihood that less benign aspects of cultural diffusion will increase and be felt in the host country.

Since we have already addressed the economic reasons for moving into large cities, let us take a look at the step-by-step process of migration, which, as evidenced by the growth in the numbers of Muslims living in the United States, is already occurring. In this "law" we can expect that Muslim families and individuals will first relocate to European cities before making a transatlantic leap. Through social networks already established during the initial flows of immigrants into American cities, European Muslims can learn how to make the transatlantic resettlement. For decades, American universities have welcomed Muslims onto their campuses where many undertook studies in the sciences and in engineering. In places like Chicago, Illinois, and Dearborn, Michigan, large Muslim communities from various countries have established themselves and their institutions. These newly formed enclaves are socially tied to similar communities in Europe and the Middle East, making the movement of people and ideas between continents a relatively easy process.

In the final analysis of the forces behind migration, economics as a means of survival is the bait that pulls people across international boundaries. The hook that keeps them in the host country is the comfort level they experience while in residence. Ethnic enclaves provide immigrants with a way to minimize the shock of long-distance migrations. In such places, immigrants see, feel, and smell the old country. If the immigrant is able to find an economic niche, she is not likely to return to the home country.

CARRYING CAPACITY IN THE NEAR EAST, MIDDLE EAST, AND SOUTHERN ASIA

It is reasonable to wonder why people who hold the opinion that the ground on which they trod is holy would want to leave such a place. Even during Muhammad's time, it was understood that life in the dry desert region was harsh. It is not surprising then that the Qur'an has much to say about environs and migration. The Qur'an certainly does not prohibit migrating for economic reasons. On the contrary, it encourages it. As was shown earlier in the chapter, Muslims who migrate for economic or political reasons are assured in their holy book that they will find that the earth contains many places of refuge that are blessed with more than adequate resources.[23] In case an immigrant were to think that his migration would be temporary, the Qur'an tells him that "the first of the migrants from Makkah [Mecca] and the people from Medina, and those who followed them in goodness, Allah is well pleased with them and they are well pleased with Him, and He has prepared for them a garden beneath which rivers flow, to abide in them for ever; that is a mighty achievement."[24] As with Jews and Christian before them, the God of Abraham has provided whatever bounty there may be across the earth as blessings for the believer's benefit and enjoyment.[25] No doubt Paleolithic and Mesolithic peoples held similar views in regard to their nature gods. Historically, a believer in any of the Abrahamic religions is not discouraged by the lack of resources and opportunity in his home country. He can move to more lush environs.

Biblical and Qur'anic concepts like bounty, flowing rivers, and gardens suggest attributes of a place well suited to sustain human life in an agrarian society. What would happen to a people when their environment is no longer able to support the level of population living in it? Wilbur Zelinsky, an American geographer, attempted to answer that important question in his 1971 article on mobility transition.[26] He logically concluded that migration would take some of the pressure off of the local environment as people leave in search of a means of survival. A little over one decade later, a Scottish geographer named Huw Jones refined Wilbur Zelinsky's model by infusing it with a loose conflict perspective.[27] Jones argued that a modern society has a number of employment sectors and, invariably, one may become saturated with too many workers while other employment sectors may have unfilled positions. Such a situation creates a surplus of labor specific to a segment of the economy and, as a result, a number of people shift about in search of suitable employment. Jones argues that these are the most likely people to move away in order to fill similar occupations in other places. It should also be pointed out that people with specialized skills or advanced educational levels are more likely to move farther from home to find suitable work. This helps explain why many Asian and Middle Eastern immigrants in the United States are generally well educated and fill highly technical positions. On the other hand, Mexican immigrants, whose home country is right next door, tend to fill low-wage jobs.

At this point, it would be helpful to insert the concept of carrying capacity into Jones's model to more fully explain and refine his views. The concept of carrying capacity is borrowed from ecology, but, with respect to humans who modify the environment, we need to consider economic structure. A robust, diversified economy is capable of absorbing many more people than an economy based on one or a few industries. Let us consider what could happen to a country that has an economy based primarily on tourism and complimentary industries like food services and hospitality if a major energy crisis were to occur. Clearly, it would seriously dampen the flow of visitors whose purchases and rental payments keep the economy running smoothly. Such an event could cost jobs and create social instability in many areas in Europe and North America. Likewise if Middle Eastern, oil-rich countries have surplus populations relative to their rather limited employment sectors, they run the risk of political upheavals. Revolutions are much more likely if large numbers of young adults and adolescents are idle. The riots that rocked Egypt, Libya, and Syria in 2011 were symptomatic of the problems that result from having too few opportunities for large youthful populations. Hence, it behooves European, North American, and oil-producing states in the Middle East to allow and, indeed, to encourage population flows between them, especially if they take pressure off the Middle East while filling low-wage service sector occupations in Europe with Middle Eastern and Asian workers. Here we can see that it is somewhat functional for this exchange of human beings to take place. It shifts population away from places where the carrying capacity is restricted and moves surplus people into places where there is an economic niche for them to fill. However, this "pressure release valve" analogy does not address the cultural and social conflicts that may result from forming large and dissimilar ethnic enclaves in foreign countries.[28] As we will see in the next section of this chapter, the demographic profiles of Near Eastern, North African, and southern Asian countries, combined with declining populations in Europe, suggest that the intensity of migration flows into Europe and North America is likely to increase over the next few decades. Latin American population figures will be placed in the mix later in this book, but for now let us take a longer look at what is going on demographically in the Eastern Hemisphere.

Before proceeding, it is important to briefly point out that one possible explanation for the civil unrest that has characterized the so-called Arab sping in the Middle East is the weakening of the economies in several nearby European countries. With fewer jobs available for guest workers, Greece, Italy, and Spain, for example, cannot absorb as many immigrants as they could during more economically vibrant times. With few employment opportunities available immediately outside of their own countries, young Syrian and Egyptian people, and indeed others from elsewhere who cannot find an ecomonic niche at home, have taken to the streets in protest. It is unclear as to whether or not this economic situation will continue. If it does, one could expect greater and more frequent social crises in the Middle East, where it remains to be seen if secularizing or geotheological forces will prevail in shaping the region's political landscape (the topic of economics throughout the eastern hemisphere is explored further in the next section). As austerity measures across Europe show, some areas already have surplus labor supplies, and others are quickly filling up. This process is likely to continue because, despite having higher gross domestic products (GDPs) and larger carrying capacities, there are limits in these areas' capacities to absorb more immigrants. This evolving situation perhaps exacerbates the fear that social conflict will result from competition for limited resources. In modern societies, carrying capacity is a function of economic and political structures, which, as Max Weber and Karl Marx pointed out, are connected to other institutions, tradition, and deeper thought worlds, including religiously inspired images of spaces and the forces of nature.[29]

Now that we have looked at the concept of carrying capacity and the forces underlying most international migration decisions, it is possible to make better sense out of the data reflecting demographic transitions in Near Eastern, North African, and southern Asian countries. Projections for the year 2050 show that most of Europe will experience negative population growth, creating a population vacuum that is sure to pull in more immigrants.[30] It is unclear what will happen to the natural increase rate among immigrants who will arrive in Europe from less developed countries, but we can expect that a reduction in their growth rates will take at least a few generations, if at all, to occur. If Ravenstein's step-by-step law has relevance, a similar type of growth will soon occur in North America. Indeed, as of 2011, it is already happening.[31] These movements of people are predicated on sustained population growth in places that have limited capacity to absorb more people. It is difficult to know at what point population in immigrant enclaves in European and North American urban areas will exceed their carrying capacities, but, with no end in sight to the current natural increase rate in North African, Near Eastern, and southern Asian countries (see table 11.1), there can be little doubt that such will be the case into the future.

Table 11.1: Demographic and Economic Profiles of Selected Countries in the Eastern Hemisphere

Country (N = 61)	2009 Population	2009 Population Density (people/sq. mile)	2009 Doubling Time (years) or Rate of Annual Decline	2009 Literacy Rate (%)	2009 Per Capita GDP (US Dollars)
Afghanistan	32,738,376	131	27	29	1,000
Albania	3,619,778	342	72	99	6,300
Algeria	33,769,669	36	58	75.4	6,500
Austria	8,205,533	257.8	Declining at -.12	98	38,400
Azerbaijan	7,911,974	237	73	99.4	7.700
Bahrain	718,306	2,797	55	89	32,100
Bangladesh	153,546,901	2,969	34	53.5	1,300
Belgium	10,403,951	890	Declining at -.02	99	35,300
Brunei	381,371	187.4	48	94.9	51,000
Burkina-Faso	15,264,735	144.4	23	28.7	1,300
Chad	10,111,337	20.8	24	25.7	1,700
Comoros	731,775	873.4	26	75.1	1,100
Djibouti	506,221	57.1	37	67.9	2,300
Egypt	81,713,517	212.6	42	72	5,500
France	64,067,790	259.2	1,800	100	35,300
Gambia	1,735,464	449.5	27	40.1	1,300
Germany	82,369,548	610.9	Declining at -.26	99	34,200
Greece	10,722,816	212.3	Declining at -.09	97.1	29,200
Guinea	9,806,509	103.3	27	29.5	1,100
India	1,147,995,898	1,000	45.6	66	2,700
Indonesia	237,512,355	336.8	55	91.4	3,700
Iran	65,875,223	104.3	64	84.7	10,600
Iraq	28,221,181	169.1	28	74.1	3,600
Italy	58,145,321	512.2	Declining at -.23	98.9	30,400
Japan	127,288,419	879.7	Declining at -.14	99	33,600
Jordan	6,198,677	174.6	41	93.1	4,900
Kazakhstan	15,340,533	14.9	102	99.6	11,100
Korea (South)	48,379,392	1,276.1	211.76	97.9	24,800
Kuwait	2,596,799	377.4	37	93.9	39,300
Kyrgyzstan	5,356,869	72.5	44	99.3	2,000
Lebanon	3,971,941	1005.6	62	87.4	11,300
Libya	6,173,579	9.1	32	86.8	12,300
Malaysia	25,274,133	199.2	41	91.9	13,300
Maldives	385,925	3,331.8	64	97.0	4,600
Mali	12,324,029	26.2	22	23.3	1,000
Mauritania	3,364,940	8.5	25	55.8	2,000
Morocco	34,343,219	199.3	46	55.6	4,100
Myanmar	47,758,181	188.1	90	89.9	1,900
Niger	13,272,679	27.1	24	30.4	700
Nigeria	146,255,306	415.9	35	72	2,000
Norway	4,644,457	39.1	400	100	53,000
Oman	3,311,640	40.4	23	84.4	24,000
Pakistan	172,800,051	574.7	35	54.9	2,600
Philippines	96,061,683	834.4	33.8	93.4	3,400
Poland	38,500.696	327.5	0 growth	99.3	16,300
Qatar	824,789	186.4	55	90.2	80,900
Saudi Arabia	28,146,657	33.9	27	85	23,200
Senegal	12,853,259	174.3	28	42.6	1,700
Sierra Leone	6,294,774	227.6	32	38.1	700
Sudan	40,218,455	43.8	35	60.9	2,200
Sweden	9,045,389	57	Declining at -.01	99	36,500
Syria	19,747,586	277.9	33	83.1	4,500
Tajikistan	7,211,884	130.9	36	99.6	1,800
Tanzania	40,213,162	117.6	32	72.3	1,300
Thailand	65,493,298	331.5	112.5	94.2	7,900
Tunisia	10,383,577	173.1	70	77.7	7,500
Turkey	71,892,807	241.6	71	88.7	12,900
Turkmenistan	5,179,571	27.5	38	99.5	5,200
U. A. Emirates	4,621,399	143.2	52	90.4	37,300
Uzbekistan	27,345,026	166.5	57	96.9	2,300
Yemen	20,013,376	112.9	21	58.9	2,300

The numbers in table 11.1 show some interesting patterns. While Albania, Azerbaijan, Kazakhstan, Lebanon, Tunisia, and Turkey require little more than seven decades for their populations to double, Afghanistan will double its population in only twenty-seven years, despite its rather protracted entanglements with relatively developed countries. These anomalies can be at least partially explained. In the case of Kazakhstan, its recent history as part of the Soviet Union surely impacted its attitudes and infrastructure to support nonreligious social institutions. Contact with the highly secular West, especially in the tourism industry, no doubt influences fertility in Tunisia. In the case of Maldives, it draws low-wage people from neighboring countries to fill positions in its vitally important tourism industry.[33] Beyond these considerations, though, much can be said about economic and environmental stresses on all of these countries.

Being situated in the Sahara Desert region of North Africa, one would not expect to see a high population density in Libya, Algeria, and Sudan, so their current population data are not too surprising. In contrast, Egypt claims 212 people per square mile, a figure that in just forty-two years will climb to 424. The human resource potential is great in Egypt, but with less than 75 percent literacy rate in its highly clustered and poor population living in the Nile corridor, residents who cannot find a niche will move away. The situation is even worse in Bangladesh. In less than thirty-five years, that country will have nearly 6,000 people per square mile. With a mostly impoverished and illiterate population now, it will only get worse in the next three decades and no doubt beyond. To give this data some scope, consider that North American cities like Cleveland and Cincinnati have smaller population densities. The population density in Maldives is already similar to Madison, Wisconsin. By 2031, Comoros will have a population density greater than Nashville, Tennessee, or Louisville, Kentucky. Also in just thirty-five years, Pakistan, the second most populous Muslim nation, will have a similar population density. Even Lebanon, with its relatively low population growth, is as dense as Jacksonville, Florida, and is already denser than Lexington, Kentucky. With an annual per capita GDP of only $11,300, the average person in Lebanon, which as a nation has historically attracted a sizable wealthy tourist population from nearby Europe and the

United States, must sense a great deal of relative deprivation. The situation there is likely to get worse because of the government takeover in 2011 by the Iranian supported Hezbollah, a Shia controlled party. Keep in mind that when we exclude oil-producing nations and former Soviet countries, data for GDP and literacy, which are critical factors in calculating a Human Development Index (HDI) as a measure of quality of life, suggest that populations in these countries are already stressing their carrying capacity.[34]

Those countries that King Abdullah of Jordan included in the Shia Crescent are also growing faster than Western nations. Data are not readily available on the Hezbollah community in Lebanon, however. As for Syria, Iraq, and Iran, each will see increases in population. By the mid-2030s, Iraq will exceed fifty million with a population density of over three hundred people per square mile. With only seven out of ten people able to read and write, the sectarian violence there is perhaps a symptom of a larger conflict over basic survival needs and which group has control over the country's limited and nonrenewable resources. Aside from the possibility of developing more efficient uses of its oil reserves and the introduction of new industries, the beleaguered desert land will be hard pressed to expand its carrying capacity. Iran will have about ninety million people by midcentury, and Syria will need to feed nearly forty million people by 2050. By midcentury, the Shia Crescent will serve as the home of approximately 180 million people. That means the desert culture area must support over two hundred people per square mile, which is twenty-eight more people per square mile than the total now living in Michigan.

Despite their oil wealth, per capita GDP is only $10,600 and $3,600 respectively for Iran and Iraq. Syria is in the middle with $4,500. There is of course no way of knowing how well national wealth is distributed, but given the squalor in places like Sadr City, which was built in 1959 to house some of Iraq's growing Shiite population, wealth distribution indeed favored Sunnis. However, as a minority group, Sunnis face an uncertain future as democracy empowers the formerly oppressed Shia. Sectarian violence will likely resurface and create an increase in the number of Sunnis immigrating to American and European cities. As Sunnis are generally better educated and more affluent than their Shia neighbors, Iraq's beleaguered economy will likely decline even more as the country loses its cultural capital.

Other Middle Eastern countries like Bahrain, Brunei, Kuwait, Qatar, and the United Arab Emirates, the so-called super oil-rich nations, have impressive per capita GDPs. However, in terms of literacy levels, they are substantially below their former Soviet counterparts that boast a literacy rate nearly 10 percent higher. Data suggest that these countries either do not value formalized education or have an inefficient wealth distribution system in place. Regardless of the reason, there are clearly large segments of people living in these countries that do not share in their respective nations' bounties. These folk, of course, provide base populations for potential immigrants who practice more traditional ways of life. No doubt religion will play a role in their interpretations of reality.

High population growth countries like Bangladesh, Nigeria, Saudi Arabia, Senegal, Sudan, and Yemen are economically much worse off than the poorest super-rich countries located in the Middle East. Thanks in large measure to the diffusion of medical technology into those areas after WWII, their current combined population of 401,033,954 will reach one billion before the decade of 2050 is over. Their current GDP per capita ranges between $1,300 and $23,200 per year. In comparison to those who live in Western countries, the average citizen of a typical Muslim country lives in exceedingly poor conditions.

THE EAST'S EXPANDED CARRYING CAPACITY AND THE DECLINING PULL OF THE WEST

Most of Europe is wealthy when compared to most Asian and African countries, although Brunei, Qatar, Kuwait, and the United Arab Emirates boast similar GDPs to moderately wealthy European nations. Saudi Arabia beats beleaguered Russia in per capita GDP, which is just under $15,000. With an annual per capita GDP of $16,300, Poland, despite its Communist past, comes closer to matching Saudi Arabia's economic output.[35] Moving westward across Europe, coastal countries with democratic and capitalist traditions are much more productive. French and German workers benefit from $33,200 and $34,200 GDPs, respectively. Similarly, Italians earn $30,400, and at $37,400, Danes produce an even higher personal income.

The Danish per capita income is still some $3,700 per year lower than that of the Swiss. The per capita GDP in the United Kingdom is $35,100. It should not be surprising then that many immigrants will be attracted to European cities. The same thing can be said of cities in the Western Hemisphere. In North America, for example, Canadians currently earn just over $38,400 per year while citizens of the United States enjoy an average annual income of over $45,800. For the time being at least, human migration from the impoverished and over-stressed parts of the ancient ecumene will continue to build in the West. However, if economic expansion continues in the East, migration flows may well shift eastward.

The rise in manufacturing and capital goods production in the East, which has the capacity to employ workers with little or no formal education, provides clear evidence that Asia is attracting investments from Japanese and Western sources. The paradox here is that in the past, incomes attracted workers while today, investment monies are attracted to places around the globe that are politically stable and relatively unencumbered by business-restricting regulations.[36] Perhaps most importantly, investors are attracted to countries that also offer low labor and shipping costs.[37] Coastal areas and rivers with outlets into oceanic shipping lanes are especially attractive because they offer opportunities to ship bulky items at the lowest cost. Asia's reemergence as a major player on the world's economic stage has been occurring at least since 1950. At that time, Western countries generated some 56 percent of all global incomes while Asia, despite its large population, earned only 19 percent of world income. By the early 1990s, Asia's share of world incomes had risen to 33 percent, and its global dominance in the twenty-first century is expected to resemble that which it enjoyed in 1820 when Asians controlled 58 percent of global earnings.[38] As predicted by financial planners and investment brokers Monty Guild and Tony Danaher in October 2010, India, Indonesia, Malaysia, Thailand, Singapore, and China will receive new infusions of liquidity. It will come directly from foreign investments: "Their economies will grow strongly as new capital flows to their growing markets from North America, Europe and other parts of Asia. This new capital will increase employment, liquidity and put some pressure on inflation. Most importantly, it will lead to increases in the stock market and real estate valuations in these countries. . . . Their economies will grow strongly as new capital flows to their growing markets from North America, Europe and other parts of Asia."[39]

It is just a matter of time before wages in Asian labor markets attract immigrants from afar, and when that happens migration flows into Europe will likely decline as immigrants from Africa, the Middle East, and south Asia seek work opportunities in the robust economies of south and east Asia. Given that economically robust coastal areas in Asia are located in earthquake, tsunami, and hurricane zones, there will be some record-breaking consequences if this scenario develops.

LIVING IN THE EASTERN HAZARD ZONE

It is tempting to embark on a discussion about the social and cultural impacts these demographic changes will have on the world, especially those that will likely affect Europe and the United States. However, because I gave considerable attention to those scenarios in *Puritan Islam: The Geoexpansion of the Muslim World*, this section instead looks at the possibility of massive losses of life that may result from highly unpredictable forces of nature. Since killer earthquakes and hurricanes are not likely to take a permanent hiatus, the only way populations will be safe from them is to adapt or avoid living in hazard zones. Because these lands are already the homes of billions of people, the latter option is not realistic.

From an air quality perspective, Asia's situation, which already poses serious health problems, is likely to worsen as more low-wage manufacturing jobs spew aerosols and gases into the atmosphere. Nine of the world's major cities located in Asia have the not-so-distinguished designation of having the highest levels of total suspended particulates (TSP) hovering in the stagnant air over them. Breathing air chock-full of particulates exacerbates asthma and other respiratory illnesses. Local transportation produces 70 to 80 percent of municipal air pollution.[40] Countries undergoing rapid industrialization such as China depend heavily on cheap and oftentimes dirty coal (such as the high sulfur–containing sub-bituminous variety) to power their tools of production.

Because of its location on a low-lying alluvial valley on the Indochina peninsula, Bangkok is of particular interest. The city of 6.7 million people sits only a few meters above sea level. Despite its proximity to hurricane alley or typhoon alley, the UN's Economic and Social Commission for Asia and the Pacific has determined that air pollution is the city's most pressing environmental issue.[41] This is an impressive declaration when one considers that Bangkok sits precariously on the Chao Phraya River plain about thirty miles north of the Bay of Bangkok, the northern-most part of the shallow Gulf of Thailand. Just as Bangladesh sits on the north side of the Bay of Bengal and New Orleans is situated on the north side of the Gulf of Mexico, the Gulf of Thailand is the northern-most section of the South China Sea. Even without typhoons, the city copes with annual flooding during the wet season of May through October. During the drier months, when high-pressure centers prevail over the Bay of Bangkok, stagnant air creates serious air quality problems. One person out of six suffers from air pollution–related illnesses. In response, the city has enacted a Green Fleets program to reduce municipal vehicle emissions. The World Bank is funding the Bangkok Air Management Project, and the European Union is sponsoring the "Asia Urbs" program.[42] While those projects are worthy, there seems to be little thought and preparation for dealing with a catastrophic typhoon. Fortunately, Thailand's population is growing slower than other south Asian countries, including those situated on the dangerous shores of the Bay of Bengal.

Tsunamis and cyclones have killed hundreds of thousands of people in Bangladesh over the last century. Even with those environmental pressures, the country has a population density of 2,900 people per square mile. In just thirty-four years, however, there will be over 5,800 people per square mile living precariously on the north end of the Bay of Bengal. India, which will soon be the world's most populous country, will have 2,000 people per square mile in only 45.6 years. These data suggest that south Asia is positioning itself for massive loss of human life. A devastating environmental crisis could generate a human tsunami with waves of people flowing into already densely populated places like Thailand and Indonesia, and those places are not out of harm's way.

SUMMARY

Western countries have a relatively high standard of living, which has been paid for by reducing births and empowering women. While empowered women in more developed countries enjoy better educations and employment choices, often passing the benefits on to their husbands or significant others as well as themselves, the forgoing or opting out of having children actually makes their societies attractive to surplus populations growing elsewhere. With fewer children being born to Western mothers, there will be a smaller pool of workers to provide services and generate tax revenue to support an increasingly elderly population. In an Eastern Hemispheric sense, however, it does little to encourage or force highly fertile societies to lower their birth rates. Economically robust places have historically served as "pressure release valves" for overpopulated regions. As long as economically rich countries attract surplus populations from elsewhere, there will be no incentive for high-growth nations to encourage lower fertility rates brought about by cultural and social change. Women's empowerment is critical for these changes to occur. However, changes in global investments that favor Asia can have a demographic ripple effect in a host of countries. For instance, by expanding their carrying capacities via economic diversification, migration flows that have fed labor into Europe may well be diverted to south and east Asia. As they have since the emergence of the earliest *Homo sapiens*, people, at least for the next few decades, will continue to migrate out of Africa and the Near East and settle in to places that continue to serve as stepping stones of diffusion or as the last bastions of culture.

While Europe's manufacturing base is giving way to less regulated Asian countries with cheap labor supplies, employment in countries like Thailand and Bangladesh presents a host of potential catastrophic human-environmental interactions. Prevailing high-pressure centers that settle in over the southern reaches of Asia from November to April have the capacity to create atmospheric conditions that can make the death toll from Donora, Pennsylvania's air pollution disaster in 1948 seem minor. Even with successful air pollution abatement programs, the increasingly dense population along the shores of the Bay of Bengal and the South China Sea is tantamount to lining up ducks in a shooting gallery. At some time in the future, a major earthquake or a powerful hurricane will strike those places and the death toll could exceed one million.

Chapter 12

THE AMERICAS IN 2060

INTRODUCTION

Imagine that the year is 2060, and an urban strip with fast-food outlets, motels, and stores selling beach paraphernalia to vacationers stretches from Corpus Christi to Galveston. The land between Houston and Galveston, too, is heavily settled by retirees and service sector workers. The same kind of sprawling development has occurred along the gulf coasts of Florida, Alabama, Mississippi, and Louisiana. With the continued collapse of the manufacturing economy in the Rust Belt, tourism- and recreation-based economies of coastal Southern states have attracted millions of permanent and seasonal residents. Urbanization also marks the banks of the Ohio and Mississippi Rivers. Opulent homes, hotels, casinos, and other new structures have been built on modestly constructed earthen mounds in flood plains along those rivers. Historical memories of the hurricanes that swamped Galveston in 1900 and New Orleans in 2005 are faded relics of the past. This is a new age, and even the 1812 earthquake that created Reelfoot Lake in Tennessee has been dismissed by the notion that improvements in technology will keep investors and residents safe from the forces of nature. Aside from some improvements in technology, there is little else that will separate these folk from their ancient ancestors who also found that natural utopias along water bodies made life easier.

Now let us suppose that within a five-year span of time (2060–2065), a hurricane comparable in size to the one that washed away Galveston in 1900 forms over the Gulf of Mexico, and the seismic tensions that have been building for two and a half centuries along the New Madrid fault line are violently released. The loss of life would be appalling to say the least, and the cash-strapped federal government of the United States would not be able to pay for the financial devastation those events would cause. By 2060, changes in global investment patterns have left few countries outside east Asia with a surplus of money to absorb a major natural disaster, let alone a series of them.

In a manner that to some observers seems as though nature is punishing sinful humanity for its exploitive behaviors, major quakes strike Japan, Chile, and Haiti, and a massive cyclone that forms over the steaming waters of the Bay of Bengal pushes a storm surge in excess of twenty feet high inland over the low coastal plain that defines the landscape of Bangladesh. In 2060, the vulnerable nation has a population of more than three hundred million. Its population density of five thousand people per square mile exceeds that of the 2011 statistics for the cities of Atlanta, Cincinnati, or Memphis. Making matters even worse for humanity, the earth's Holocene, the interglacial warm period that some call the Age of Man, nears its end. Glacial ice is advancing in heretofore green mountain valleys. North Atlantic shipping lanes are clogged with ice sheets that get thicker each year, making heavy-goods shipping more costly as routes get longer. The loss of life and the impact that these events could have on the global economy would be unimaginable and unsustainable. While the events described here are imaginary, they could happen. Indeed, they will happen, but it is impossible to know when and in what frequency they will occur. If such a series of events happens, how will countries respond? Will they be able to respond? Those are difficult questions to answer with a high degree of certainty.

As the last earthly frontier to be settled by the descendants of a man who lived in a valley in Ethiopia some sixty thousand years ago, the Americas may not offer much hope, for despite the vast prairies that stretch across North America's open interior, that particular region is extremely dangerous. South America's Amazon River watershed that covers most of the northern reaches of the continent also poses serious environmental challenges to any society, if not the world, that wishes to deforest and urbanize the region's vast wetlands. As we look ahead to 2060, it is arguably best to apply what we know about human settlement patterns and evolving transportation arteries to create a future, albeit imagined, geography of the Americas.

There are currently just under nine hundred million people living in the Americas and on the larger islands of the Caribbean Sea. In just one person's lifespan, about seventy-eight years, that number will double to about 1.8 billion humans. Shoreline areas will see an unprecedented increase in population density as will inland places where highways are built to connect far distant coastal cities. Not only will coastlines bulge with human occupants, it is conceivable that interstate highways like I-44, I-35, and I-70 that crisscross the heartland of the United States will serve as the skeleton for urban areas springing up in heretofore sparsely populated parts of the United States. No doubt the Mississippi River Valley will also experience an increase in population. Such a situation will place millions of people in harm's way. Suppose, for instance, that a megalopolis forms along Interstate 35 between Dallas, Texas, and Oklahoma City, Oklahoma. The death toll from an F5 tornado like the one that devastated this section of the country on May 3, 1999, would be catastrophic.

Admittedly, however, global shifts in investments and an accompanying decline in the North American manufacturing economy make it difficult to speculate on the United States's role in the Western Hemisphere in 2060. It also makes it somewhat challenging to accurately predict how population change will unfold in the United States. In the past, the country's high standard of living has attracted millions of immigrants from Latin America as well as from Africa, Asia, and Europe. If the US economy does not rebound from the recession of 2009–2012, it is likely that the country's attractiveness to potential immigrants will decline, although various social net programs will continue to attract some immigrants. However, without a tax base to feed it, the future of that economic niche is also in doubt. This is especially true when one recognizes that the federal government's $14 trillion deficit has been incurred to maintain Americans' high living status. Without a robust economy, the current fiscal situation is not sustainable in the United States.

It is no easier to speculate on the culturally sensitive carrying capacity of Central and South America. As we discussed in chapter 11, those important parts of the Americas are experiencing some inflow of foreign investments, including moneys that in the recent past would have been invested in the United States. However, as we saw in chapter 3, Central and South America do not offer ideal natural settings and climate conditions for sustainable human development. While it is not easy to imagine how societies in the Americas will function at the end of the middle decade of the twenty-first century, there is little reason to believe that current demographic patterns among them will change. On the other hand, without radical changes in tectonic and atmospheric forces that have visited the Americas in the past, it is possible to make some cautionary predictions of unsavory environmental situations that will likely unfold in them.

LANDS OF CONTRASTING ENVIRONMENTAL FEATURES

Much of Central America, including the narrow isthmus that connects the region to South America, is hot and humid. Fragile rainforests, whose very presence help produce some of the most biologically diverse ecosystems on the planet as well as our vital supply of oxygen, flourish in the lands south of Mexico. Like their counterparts in South America, regional tree species such as roble de sabana (Savannah oak) and magnolia poasana help supply the entire planet with oxygen while down-cycling huge quantities of CO_2. The aesthetic qualities of the region attract visitors from more temperate climates. Ecotourism is, hence, emerging as a viable economic enterprise in many parts of Latin America. However, ecotourism is not compatible with large, densely populated urban areas, nor is it compatible with rural areas in which deforestation and impractical forms of agriculture are practiced. In other words, ecotourism is not a viable industry for societies in which large numbers of low-skill, low-wage workers can be employed. Unfortunately, providing work opportunities for growing populations is essential in creating economic and social conditions conducive to lowering a society's natural increase rate. In other words, work tied to traditional extractive industries such as farming and mining depend heavily on male workers, allowing women to care for the home and rear children. A more diversified economy with a substantial manufacturing sector requires male and female workers, so having children in tow is not practical for women employed outside the home. In such cases, children are no longer seen as assets (such as sources of labor on farms, as well as social security providers for elderly relatives). Indeed, they become financial liabilities. As a result, population growth rates, as we have seen in Japan, Germany, and Italy, for instance, steadily decline. The paradox is that while ecotourism is good for the environment, it is not compatible with an economy that leads to lower population growth.

South America, too, is an environmentally fragile place. Along the west coast lies the Atacama Desert. Inland, the Andes Mountains rise abruptly from the desert floor to reach an elevation of 22,834 feet (6,960 meters) at Aconcagua in Argentina. Those South American mountains form a spine of steep-sloped, rocky earth that divides the arid west from the moist, biologically diverse interior of the continent. Ancient alpine glaciers still carve away at exposed rock near the summit in the northern, equatorial reaches of the Andes and at the lower elevations in the colder south. Much of the eastern Andes and the northern interior of South America form the watershed for the Amazon River. This portion of the continent is rainy throughout the year as daily doses of energy from a nearly vertical sun produce a persistent supply of convective rain events. The rugged uplands of the southern Andes converge with the grasslands of Patagonia to form the southern third of the continent. Here the landmass forms a narrow cape that points toward Antarctica, so there is precious little land situated in a temperate climate zone. Unfortunately, such a climate best suits people. To escape the oppressive heat and excessive wetness of the northern interior of South America, most of the population of the continent is pressed along the eastern coastline where the moderating influences of the Atlantic Ocean make life there more bearable.

In the Western Hemisphere, North America offers the largest land area that lies in a temperate climate zone, which roughly parallels the midlatitudes (30° north to 60° north). Although not particularly a fragile place like the Amazon rainforest, the interior poses its own environmental hazards and limitations, including Tornado Alley and little surface water. California has its earthquake-prone San Andreas Fault and its life-limiting, semiarid interior where irrigation creates salinization as, over time, rapidly drying fresh water leaves behind traces of salt that accumulate; the interior western states are marked by cold uplands and dry valleys; the Eastern Seaboard is already densely populated. North America's colliding masses of dry-cold and warm-moist air will pose greater threats to human life as population densities increase in its interior.

To cast more light on a possible future geography of the Americas, it is necessary to shift our attention to demographic transitions that are occurring among twenty-nine of the largest countries in the Americas and the Caribbean. While imagination informed by knowledge can help us see where danger lies, only future generations will know how they will adapt to the West's own formidable forces of nature.

POPULATION CHANGE IN THE AMERICAS

As the last hemisphere to experience human settlement, the West is now the host to every racial group on the planet. Since the arrival of Columbus in 1492, the population of the Americas, including the Caribbean nations, has climbed from fewer than one hundred million, which is a highly speculative estimate, to over 880 million people.[1] However, in less than fifty years the population will grow well beyond one billion. While the Americas are the homes of all races of people, there are spatial patterns among them. There are also patterns of economic development that are quite contrastive. Indeed, the range of per capita GDPs among American countries is more than $40,000 per year (see table 12.1 on page 271). However, this rather large range is subsidized heavily by the United States, which is borrowing significant sums of money from China and Japan. In other words, the real range is closer to $35,000, which is still an impressive figure. On the other end of the economic spectrum is Haiti. Its per capita GDP of $1,300 places it among the poorest countries in the world. In less than thirty years, its population density will match or exceed a number of cities in the United States. Mexico will reach a population of over two hundred million people by 2060. Given the flow of people from that country as well as from Haiti,

Table 12.1. Doubling Times for Countries in the Americas (N = 29)

Country[2]	2009 Population	Doubling Time (years)	Population Density (people/sq. mile)	Per Capita GDP (USD)	Literacy Rate (%)
Antigua and Barbuda	84,522	68	495	18,300	85.8
Argentina	40,481,998	67	38	13,300	97.6
Bahamas	307,451	92	79	25,000	95.6
Barbados	281,968	185	1,694	19,300	99.7
Belize	301,270	33	34	7,900	70.3
Bolivia	9,247,816	48	22	4,000	90.3
Brazil	196,342,587	58	60	9,700	90.5
Canada	33,212,696	267	9	38,400	99
Chile	16,454,143	79	56	13,900	96.5
Colombia	45,013,674	50	112	6,700	93.6
Costa Rica	4,195,914	54	214	10,300	95.9
Cuba	11,423,952	176	266	4,500	99.8
Dominican Republic	9,507,133	41	509	7,000	89.1
Ecuador	13,927,650	42	130	7,200	92.6
El Salvador	7,066,403	36	883	5,800	85.5
Grenada	90,343	47	680	10,500	96
Guatemala	13,002,206	31	311	4,700	73.2
Guyana	779,794	75	10	3,800	98.8
Haiti	8,924,553	28	839	1,300	62.1
Honduras	7,639,327	33	177	4,100	83.1
Jamaica	2,804,332	53	671	7,700	86
Mexico	109,955,400	47	148	12,800	92.4
Nicaragua	5,785,846	37	125	2,600	80.5
Panama	3,309,679	45	113	10,300	93.4
Paraguay	6,831,306	30	45	4,500	93.7
Peru	29,180,899	53	59	7,800	90.5
Suriname	475,966	63	8	7,800	90.4
Trinidad and Tobago	1,047,366	313	529	18,300	98.7
United States	303,824,646	122	86	45,800	99
Totals and Averages	881,500,840	78	N/A	11,493	N/A
Adjusted Averages*	N/A	50	N/A	9,226	N/A

*Adjusted per capita GDP is without the USA and Canada. Adjusted doubling time is without the USA, Canada, Trinidad and Tobago, Cuba, and Barbados. Doubling time is calculated by the Rule of 72. To calculate the doubling time, 72 is divided by the natural increase rate (NIR).

it is clear that their carrying capacities have been exceeded. One can only imagine the plight of the Mexican and Haitian peoples if they had twice as many mouths to feed. However, if their current fertility rates spread with them, other places, too, will become overpopulated.

Carrying capacities will also be exceeded in other Latin American countries. Situated between Cuba and Puerto Rico, the island of Hispaniola is one of the most densely populated islands in the Americas. Two sovereign countries share the island. Haiti occupies the western reaches of the island while the Dominican Republic governs the eastern portion. As mentioned above, Haiti's population density, which is already 839 people per square mile, will double to 1678 people per square mile in just twenty-eight years. Thirteen years later, the Dominican Republic, too, will reach the threshold of one thousand people per square mile. These countries sit astride an earthquake zone, so at some point in the future, a strong quake will strike the island. Because of its much denser human population, the loss of life on Hispaniola will be catastrophic. The loss of life could be reduced substantially by adopting the successful construction designs used by the Japanese. As a case in point, the substantial loss of life resulting from the March 2011 earthquake in Japan was caused mostly by tsunamis inundating low-lying coastal cities. The shaking and collapsing of homes and businesses caused comparatively few deaths.

While the effect of overpopulation is not readily apparent in the data contained in table 12.1, the factors associated with it are nevertheless in place. There are thirteen countries in which the doubling time is less than fifty years. Of those thirteen, only Mexico and Panama boast per capita GDPs greater than $10,000. On the other hand, Guatemala and El Salvador are more typical of Central American countries. In just thirty-one years, there will be 622 Guatemalans crammed into one square mile. Less than three-fourths of the population of Guatemala can read, so its workforce will be well suited to fill low-wage manufacturing jobs. Without external investments to expand the country's manufacturing economy, however, more of the local rainforest will be depleted for cultivation to feed its population. The situation is perhaps even more precarious in El Salvador. In just thirty-six years, that country will have more than 1,600 people per square mile. Again, without external investments in the country's meager manufacturing economy, families will have little choice but to clear more land to make way for intensive agriculture.

Lands once occupied by rainforests are not well suited for commercial agriculture due to the processes of leaching and lateralization. As water from persistent rainfall events percolate down through soil horizons, root systems of native plant species keep nutrients and moisture near the ground surface. However, in the absence of those plants, nutrients are washed or leached away from upper soil layers, making the land infertile for commercial crops. Also, leaching leaves behind laterite, a mineral that, over time, reduces soil arability while making the ground quite hard.

In South America, Brazil's population is growing a little slower than the rest of South and Central America. Nevertheless, in just sixty years, the Portuguese-speaking behemoth will be the home of four hundred million people. Its per capita GDP is less than $10,000 per year, so poverty is wide spread. If there is a bright spot for Brazil, it is in its education system. Ninety percent of the population is literate. Populations in other South American counties like Ecuador and Paraguay are growing much faster than Brazil. While Chile is not as developed as the United States or Canada, the earthquake-prone country boasts a 96.5 percent literacy rate and nearly $14,000 per capita GDP. However, much of the country is desert, so the thirty-three million people who will live there in 2090, will have to rely heavily on technology to survive.

THE FATAL ATTRACTION OF NORTH AMERICA

The issues of human adaptation and survival have been the central themes of this book. Indeed the spread of humanity around the globe was produced by the need for space to bag game and gather edibles under conditions of less competition. This was especially true for the hunters and gathers that bequeathed their genes and accumulated knowledge to us. Few people today, however, are hunters and gatherers, but the Americas still attract people from around the world. They are especially drawn to the labor markets of North America. In the United States, the number of foreign-born Americans topped thirty-three million in 2002, which was more than 11 percent of the country's population. In 2008, the US Census Bureau reported that the number of foreign-born Americans had risen to 38.5 million or about 12.5 percent of the country's 307 million people.[3] Some 55.2 percent of the foreign-born population emigrated from other parts of the Americas. This is the equivalent of all of the citizens of the Bahamas, Barbados, Belize, Costa Rica, Guyana, Jamaica, Nicaragua, and Panama pulling up stakes and moving to the United States.

It is important to look at the environmental situations of those states that attract immigrants from other countries in the Americas. Texas, California, and Florida, as well as the relatively safe environs of New York, are magnets for international folk. With the exception of New York, which will not be out of harm's way when an ice age returns, each of those states presents a number of human-environmental hazards that will only worsen if population densities in them increase. While much attention is given to the San Andreas Fault zone in California, few people outside of seismologists know that the main fault zone actually has a number of branch faults. One of those is the Hayward Fault. It extends through the east Bay Area of San Francisco. At almost seventeen thousand people per square mile, the city already has a population density twice as high as Los Angeles.[4] Seismologists estimate that a release of pressure along the Hayward Fault will generate a quake in the range of 7.0 magnitude. There is a 28 percent probability that a quake of that strength will occur by 2018.[5]

Not all quakes generate the same kind of seismic waves and, hence, the same kind of damage. If a tectonic event were to occur along the San Andreas Fault about thirty-one miles north of the downtown area of Los Angeles, the release of energy would be enormous and so would the resulting damage to infrastructure. The loss of life would be widespread and catastrophic.[6]

Although states located along the Mississippi and Ohio Rivers are not magnets for huge numbers of immigrants from the Americas, they make up a region in which the domestic population will continue to grow. The New Madrid Fault underlies a portion of the Mississippi River. In the event of another quake in the magnitude of those of 1811 and1812, the loss of life near the fault would be large. While the population density of Louisville, Kentucky, closely matches that of Los Angeles, the current densities of Memphis and St. Louis are mercifully much less. There are, however, cities in the damage zone that were not there in 1811 and 1812. Cape Girardeau, Missouri; Paducah, Kentucky; Jonesboro, Arkansas; and other communities that lie outside the boundaries of the major cities bring the population in the hazard zone to nearly two million. Given that builders in California are more aware of the need to erect flexible structures than their counterparts in the Mississippi and Ohio River Valleys, damage in California is likely to be less than that along the Mississippi and its major tributary. Based on their impacts on local landscapes, it should be pointed out that the quakes of 1811 and 1812 produced vertical and horizontal energy waves. Structural designs that can absorb those kinds of shaking are expensive. Memphis, St. Louis, and Louisville are cities in which large numbers of poor people live in brick and mortar–based housing developments. Those brittle structures may well be death traps for hundreds of thousands of Americans.

Gulf coast states are vulnerable to destructive hurricanes. As development continues along that portion of the American coastline, investments and lives will be at stake. Again, it is not a question of if a category 5 hurricane will make landfall. It is rather a question of when it will happen. Cities in Florida like Miami and Hialeah are magnets to retirees from the north. Most of those transplants do not have family members living nearby, so in the event of a powerful hurricane, they depend on government agencies to find refuge from high winds, torrential rain, and storm surges—a hurricane's most lethal weapon.

As mentioned earlier, the Great Plains, too, are dangerous places for people, especially those who are not highly mobile. This part of the country is attractive to immigrants from Latin America because without advanced educations or specialized training they can find work on ranches, on farms, or in some other agribusiness. Unfortunately for them, a lack of local knowledge of weather events could place them in jeopardy. Poverty makes it difficult for residents to live in the safest homes. Subsidized housing usually takes the form of brick and mortar structures. If they are able to find residences outside of housing developments, they typically live in mobile homes or cheaply built frame houses. Neither of these types of structures stands much of a chance against an F4 tornado. In the event that the community in which they live has a public storm shelter, not having adequate English language skills as well as reliable transportation limits them in their ability to learn about and subsequently seek refuge in the event of a major storm event.

THE NEED FOR DEVELOPMENT IN THE AMERICAS

Cultural change within a society or in comparison to another people occurs through innovation, isolation, or interaction with others. A look back across millennia shows how our ancestors carved out niches for themselves and their progeny. As we saw earlier in the book, the life of the hunter and gatherer was not willingly discarded. Such a life facilitated large amounts of leisure time. However, as spaces available to hunt and gather edibles declined through competition with others, survival needs forced ancient humans to use their accumulated knowledge of flora and fauna to begin a sedentary life of tilling the soil and domesticating animals. The locations of the earliest civilizations suggest that migrating peoples simply ran out of places to forge ahead. Even during the Mesolithic era, hunters and gatherers made rudimentary attempts to create cities at the edges of the world. In ancient places like Skara Brae, Mount Sandel, and Hemudu, hunters and gatherers reached the edge of their known worlds, looked out over the oceans before them, and were forced to make some decisions. Most likely they looked around the places where they stood and discovered that with a little effort they could lead a semipermanent life with roofs over their heads. Innovations are born in such moments of need.

While few people today are hunters and gatherers, they are nevertheless hunters of economic niches. As such, they, like their ancestors before them, are willing to cross inhospitable lands and seas to arrive at a place where they can live life at its fullest. There is an underlying problem with this situation, despite its lengthy pattern throughout human history. Countries with the highest fertility rates are not found at the edges of the world (islands and coasts of continents) unless they have a highly resilient belief system that maintains traditional gender roles as well as an outlet for sending surplus populations. A cursory look at table 11.1 (in chapter 11) and table 12.1 (within this chapter) suggests this peculiar spatial pattern. Countries like Barbados, Chile, Cuba, Guyana, Japan, Germany, Russia, and Canada, are in many ways at the edge of their worlds. There are few places for them to send excess populations, so they have had to develop other economic niches. Along the way changes took place in their gender roles, and their perception of children changed from being assets to becoming financial liabilities.

As was shown in the previous section, the United States has been a magnet for excess populations growing in other countries. While this situation looks on the surface to be a morally "good thing" for everyone involved, the situation does little to force other countries to change their ways of life. While large flows of immigrants have provided a cheap labor source for American industry in the past, other cost factors have made consumer goods produced in Asia relatively inexpensive. America's tertiary and quaternary economic sectors cannot employ large numbers of underskilled workers. The nation's debt, which continues to accumulate, is a product of subsidizing American workers to stay out of the labor market as well as financing multiple wars. Indeed, large armies, navies, and air forces keep able-bodied people out of the labor market. By restricting the supply of labor, wages are kept high.

The high cost of production in the United States and Canada may, in the long run be a good thing for the Americas. While that statement may seem insensitive to some observers, it will mean that fewer jobs will be available to low-wage workers. Since potential immigrants to the United States will not be in a financial position to make transpacific relocations, they will more than likely be forced to stay in their home countries. Internal social and political pressures will increase in highly fertile countries. Their economic reality will create a situation in which they, too, will be placed figuratively at the edge of the world, albeit the economic world. They will have to adapt or suffer one of Thomas Malthus's dire predictions: famine, disease, or warfare.

SUMMARY

As America declines in its capacity to absorb more people (as is the case as of 2011), other countries or societies will have to adapt by changing gender roles; modifying religious beliefs; and developing better early warning systems, storm shelters, and a host of other technologies in order to survive. While I am not a proponent of killing off American industry to save other peoples, it is a process that is occurring. As those countries industrialize, they will have to do so in a delicate balance with nature. To save the world's population, it is important for other countries to increase their economic output, thus changing the roles that women play in the home and workplace. This will then reduce their natural increase rates. On the other hand, if the United States stays on top of the world's economic power structure, demographic flows will continue into the country from less-developed parts of the Americas and Africa, including the Middle East. Global population will continue to climb, and that scenario is not sustainable. The larger and more difficult questions are these: How will Americans adapt to a lower standard of living? Will more traditional roles for men and women reemerge in North America as their value in the workplace lessens?

Beyond matters of basic carrying capacity, which is a function of economic structure and resource utilization, there is the need to safeguard native and immigrant populations alike from environmental hazards. Natural disasters come in many guises; some attack from across the Atlantic in the form of hurricanes while others rise up violently as pressure is released by tectonic forces below the surface of the planet. Given that the best living spaces are already densely populated, more people will be forced to live in the interiors of North and South America. Whereas developing the hot wetlands of South America in a random fashion will have a ripple effect on the planet's atmospheric conditions, increasingly dense population centers in the Great Plains and along the Mississippi River Valley will need better early warning systems for tornadoes, more flexible building designs that allow the power released by earthquakes to travel through structures, and underground storm shelters.

The future poses many challenges. With an awareness of the power of the forces of nature and the limits of our own abilities, development can proceed in South America. Green zones or national parklands in the rain forest can be set aside and protected from development. However, developments in flood plains should be limited if not completely prohibited. While levees can protect population centers from floods that emanate from surface water flow, they do little to protect lives against groundwater rise. While technology has its limits in protecting us from some natural disasters, we cannot live without it. As a product of accumulated knowledge beginning in the ancient times of early humans and reaching all the way into the twenty-first century, the developments, successes, and failures of technology all reflect how we have coped with the forces of nature.

CUMBERLAND GAP SURVEY

Information contained in this survey was presented to residents in the Appalachian region in 2010 (except for Rufus Voiles, who discussed these issues with the author in 1995).

I am conducting a survey on perceptions of how southern Appalachian families, particularly those who lived in and around the Cumberland Gap area, got along during the Great Depression. I realize that you may have been born after that time period. Nevertheless, you have families that discussed their lives with you during that important phase of American history. If that is the case with you, I would like to invite you to participate in my study. It should take less than five minutes of your time.

If you are interested in participating and your heritage is Appalachian, please answer the following six questions:

1. Did your family farm during the Great Depression (1930s)?
2. Did they have any other non-farm income? If so, what occupations or business did they perform?
3. In what county and state did your family reside?
4. Did unemployed people seek help in the form of food or work from your farm family?
5. Was your family better or worse off than city folk?
6. As this survey may result in a publication, may I cite you as a source?

Many thanks,
Barry A. Vann

ACKNOWLEDGMENTS

I t is virtually impossible to write a book of this kind without some encouragement and support from others. The idea of writing *The Forces of Nature* was planted in my mind by Dr. Frank L. "Pete" Charton during a course he delivered in atmospheric science when I was a freshman in college. Decades later, the seed that Dr. Charton planted was watered by my good friend Dr. Chin Teck Tan, who patiently listened to my impromptu ramblings about tornadoes, hurricanes, earthquakes, and how they affect human settlements. I would not have written this book without Chin's initial and continued encouragement. Amy, Sarah, and Preston Vann were also fountains of emotional support. Special thanks are due to Nick Cockrum and Lane Simmons for their photographic contributions. A deep debt of gratitude is owed to my colleague Geraldine Allen and her son-in-law Clayton Andrew Long; their maps help the book tell its story. I must also thank Debbie Wood for her assistance. Finally, I would like to offer special thanks to Steven L. Mitchell and Mariel Bard of Prometheus Books.

Finally, it is necessary to thank my students who offered me thoughtful reactions to the manuscript; among them are Hannah Adkisson, Whitney Arnold, Jeremy Bird, Madison Branstetter, Mathew Brotherton, Timber Craig, Dale Davis, Marquee Dawson, Andrew Fox, Latosha Howard, Jessica Stanfill, and Kelsey White.

NOTES

CHAPTER 1. THE POWER OF ENVIRONMENTAL PERCEPTIONS

1. Paul R. Ehrlich, *Population Bomb* (New York: Ballantine Books, 1968); Paul R. Ehrlich and Anne H. Ehrlich, *Population Explosion* (New York: Simon and Schuster, 1990).

2. Norman Maclean, *A River Runs through It, and Other Stories* (Chicago: University of Chicago Press, 1976). While some regard the book as fiction, Robert Redford considered it an autobiography and turned the book into a screenplay in 1992. The motion picture version was entitled *A River Runs through It*, directed by Robert Redford (Burbank, CA: TriStar, 1993), DVD.

3. John F. Kennedy, "America's Cup Speech," (speech, America's Cup Races, Newport, RI, September 1962).

4. Thomas Malthus, *An Essay on the Principle of Population* (Oxford: Oxford University Press, 1798). See also Adam Smith's *An Inquiry into the Nature and Causes of the Wealth of Nations* (London: Strahan and Cadell, London, 1776). Charles Darwin drew upon some of Malthus's ideas in formulating his ideas on evolution.

5. See T. C. Smout, *A History of the Scottish People 1560–1830* (London: Fontana, 1998), p. 88. Smout refers to the years after 1690 as the prelude to the take off of the Industrial Revolution. I argue that the social infrastructure for the Industrial Revolution began decades earlier in Scotland. See my work titled *In Search of Ulster-Scots Land: The Birth and Geotheological Imagings of a Transatlantic People, 1603–1703* (Columbia: University of South Carolina Press, 2008), pp. 36–38.

6. Spencer Wells, *Deep Ancestry: Inside the Genographic Project* (Washington, DC: National Geographic Society, 2007), pp. 100–104.

7. Bryan Sykes, *Saxons, Vikings, and Celts: The Genetic Roots of Britain and Ireland* (New York: W. W. Norton, 2007), p. 126.

8. David Goldfield et al., *The American Journey: A History of the United States* (Upper Saddle River, NJ: Pearson Prentice Hall, 2006), p. 13.

9. Ibid. See also Bryan Sykes, *The Seven Daughters of Eve: The Science that Reveals Our Genetic Ancestry* (London: Bantam Press, 2001).

10. James Rubenstein, *The Cultural Landscape: An Introduction to Human Geography*, 7th ed. (Upper Saddle River, NJ: Prentice Hall, 2002), p. 7.

11. See, for instance, H. C. Darby, *Relations of History and Geography: Studies in England, France, and the United States* (Exeter, UK: University of Exeter Press, 2002). See also Chris Philo, "History, Geography, and the Still Greater Mystery of Historical Geography," in *Human Geography: Society, Space, and Social Science*, ed. Derek Gregory, Ron Martin, and Graham Smith (Minneapolis: University of Minnesota Press, 1994).

12. Philo, "History, Geography, and the Still Greater Mystery of Historical Geography."

13. Alan R. H. Baker, *Geography and History: Bridging the Divide* (Cambridge, UK: Cambridge University Press, 2003).

14. Carl O. Sauer, "Foreword to Historical Geography," *Annals of the Association of American Geographers* 31, no. 1 (1941): 1–24.

15. Preston E. James and Geoffrey J. Martin, *All Possible Worlds: A History of Geographical Ideas* 2nd ed. (New York: Wiley and Sons, 1981); Rubenstein, *The Cultural Landscape*.

16. Ellsworth Huntington and Barry A. Vann, *Geography toward History: Studies in the Mediterranean Basin, Mesopotamia and Central Asia* (Piscataway, NJ: Gorgias Press, 2008), pp. 1–5.

17. Rachel Carson, *Silent Spring* (Boston: Houghton Mifflin, 1962); Aldo Leopold, *A Sand County Almanac: And Sketches Here and There* (Oxford: Oxford University Press, 1949).

18. René Dubos, *Man Adapting* (New Haven, CT: Yale University Press, 1965); Donald Carr, *The Breath of Life* (New York: W. W. Norton, 1965); William Wise, *Killer Smog* (Chicago: Rand McNally, 1968).

19. Oliver S. Owen, *Natural Resource Conservation: An Ecological Approach*, 3rd ed. (New York: Macmillan, 1980), pp. 10–15.

20. Paul Ehrlich and Ann Ehrlich, *Extinction: The Causes and Consequences of the Disappearance of Species* (New York: Random House, 1981). See also Ehrlich

and Ehrlich, *Population Explosion*; and Ehrlich and Ehrlich, *One with Nineveh: Politics, Consumption, and the Human Future* (Washington, DC: Island Press, 2004); and Ehrlich and Ehrlich, *The Dominant Animal: Human Evolution and the Environment* (Washington, DC: Island Press, 2008).

21. Huntington and Vann, *Geography toward History*.

22. See, for example, Ellen Churchill Semple, "The Anglo-Saxons of the Kentucky Mountains: A Study in Anthropogeography," *Geographical Journal* 17, no. 6 (1901): 588–623; Harry M. Caudill, "Kentucky and Wales: Was Ellen Churchill Semple Wrong?" in *Cracker Culture: Celtic Ways in the Old South*, Grady McWhiney (Tuscaloosa: University of Alabama Press, 1988).

23. Paul L. Knox and Sallie A. Marston, *Human Geography: Places and Regions in Global Context*, 4th ed. (Upper Saddle River, NJ: Prentice Hall, 2007), p. 57. There is no mention of Ellen Churchill Semple or Ellsworth Huntington in Jerome D. Fellmann, Arthur Getis, and Judith Getis, *Human Geography: Landscapes of Human Activities*, 9th ed. (Boston: McGraw Hill, 2007).

24. See Philo, "History, Geography, and the Still Greater Mystery of Historical Geography," p. 258.

25. This is quoted from page 32 of Ellsworth Huntington's article titled "Geographer and History," *Geographical Journal* 43, no. 1 (1914): 19–32.

26. Harlan Read Barrows, "Geography as Human Ecology," *Annals of the Association of American Geographers* 13, no. 1 (1923): 1–14.

27. William A. Schwab, *The Sociology of Cities* (Englewood Cliffs, NJ: Prentice Hall, 1992), p. 12; see also Harlan Barrows, *Geography as Human Ecology* (Indianapolis: Bobbs-Merrill Reprint Series in Geography, 1923). See also L. F. Schnore, "Geography and Human Ecology," *Economic Geography* 37, no. 3 (July 1961): 207–17; Philip W. Porter, "Geography as Human Ecology: A Decade of Progress in a Quarter Century," *American Behavioral Scientist* 22, no. 1 (1978): 15–39; Karl. S. Zimmerer, "Human Geography and the New Ecology: Prospect and Promise of Integration," *Annals of the Association of American Geographers* 84, no. 1 (March 1994): 108–25.

28. Baker, *Geography and History*.

29. Schwab, *Sociology of Cities*, p. 12. For a further look at the differences between the sociocultural and neoorthodox approaches to spatial perceptions, see also C. T. Jonassen, "Cultural Variables in the Ecology of an Ethnic Group," *American Sociological Review* 14, no. 5 (1949): 32–41; Walter Firey, *Land Use in Central Boston* (Cambridge, MA: Harvard University Press, 1947); Herbert J. Gans,

Urban Villagers: Group and Class in the Life of Italian Americans (New York: Free Press, 1962); and Amos Hawley, *Human Ecology: A Theory of Community Structure* (New York: Ronald Press, 1950).

30. Richard J. Blaustein, "Kudzu's Invasion into Southern United States Life and Culture," United States Department of Agriculture (2001). http://www.srs.fs .usda.gov/pubs/ja/ja_blaustein001.pdf/ (accessed May 21, 2009).

31. E. G. Bowen, "Le Pays de Galles," *Transactions and Papers of the Institute of British Geographers* 26 (1959): 1–23.

32. See John K. Wright, "Notes on Early American Geopiety," *Human Nature in Geography: Fourteen Papers, 1925–1965* (Cambridge, MA: Harvard University Press, 1966); Yi-Fu Tuan, *Morality and Imagination: Paradoxes of Progress* (Madison: University of Wisconsin Press, 1989); Avihu Zakai, *Exile and Kingdom: History and Apocalypse in the Puritan Migration to America* (Cambridge: Cambridge University Press, 2002); Vann, *In Search of Ulster-Scots Land*.

33. It is interesting that Chris Park does not offer any substantive discussion on the themes and topics advanced by Wright and even erroneously suggests that Tuan is the person associated with coining "geoteleology." See Chris Park, *Sacred Worlds: An Introduction to Geography and Religion* (New York: Routledge, 1994), p. 19. It is perhaps important to make a distinction between what R. W. Stump calls religious geography and the geography of religions. Religious geography is concerned with understanding how religious thought worlds impact human actions that may have a direct influence on space. See Roger W. Stump, "The Geography of Religion: Introduction," *Journal of Cultural Geography* 7, no. 1 (1986): 1–3. It therefore considers how the creation of imagined worlds impacts the landscape. On the other hand, the geography of religions identifies the spatial extent of religious groups (i.e., Christian denominations). Such an approach lends itself more to the creation of maps. For a more in-depth discussion on the distinctions between geography of religions and religious geography, see Vann, *In Search of Ulster-Scots Land*, pp. 10–25.

34. John K. Wright, "Terrae Incognitae: The Place of Imagination in Geography," *Annals of the Association of American Geographers* 37 (1947): 1–15.

35. Ibid.

36. William D. Pattison, "The Four Traditions of Geography," *Journal of Geography* 63, no. 4 (1964): 211–16. In addition to the man-land or human-environmental tradition, Pattison observed three other traditions or kinds of studies conducted by geographers. They include spatial, area studies, and earth science traditions.

CHAPTER 2. LIVING ON THE EDGE OF THE WORLD

1. Spencer Wells, *Deep Ancestry: Inside the Genographic Project* (Washington, DC: National Geographic, 2007).

2. Bryan Sykes, *The Seven Daughters of Eve: The Science that Reveals Our Genetic Ancestry* (London: Bantam Press, 2001), p. 278; Helen A. Gaudette and G. A. Clark in *World Almanac and Book of Facts*, ed. Sarah Janssen (New York: World Almanac Education Group, 2008), p. 658.

3. Sykes, *Seven Daughters of Eve*.

4. Ibid., p. 278. Sykes has named the mother of all non-Africans "Lara."

5. Ibid., pp. 271–86. Here I am using the time frame established by Sykes.

6. E. G. Bowen, *Britain and the Western Seaways* (London: Thames and Hudson, 1972); E. G. Bowen, *Saints, Seaways, and Settlements in Celtic Lands* (Cardiff, UK: University of Wales Press, 1969); Bryan Sykes, *Adam's Curse: A Future Without Men* (New York: W. W. Norton, 2004); Sykes, *Seven Daughters of Eve*; Sykes, *Saxons, Vikings, and Celts*; Spencer Wells's work includes *Deep Ancestry* and *The Journey of Man: A Genetic Odyssey* (Princeton, NJ: Princeton University Press, 2002).

7. Wells, *Deep Ancestry*; Sykes, *Seven Daughters of Eve*, p. 278.

8. Sykes, *Saxons, Vikings, and Celts*.

9. Marshall T. Newman, "The Blond Mandan: A Critical Review of an Old Problem," *Southwestern Journal of Anthropology* 6, no. 3 (1950): 255–72.

10. John Alexander Williams, *Appalachia: A History* (Chapel Hill: University of North Carolina Press, 2002), p. 21. Duane H. King, *The Cherokee Indian Nation: A Troubled History* (Knoxville: University of Tennessee Press, 1979), p. x. Sarah DeCapua, *The Shawnee* (Tarrytown, NY: Marshall Cavendish, 2008, pp. 7–8.

11. Simon Schama, *A History of Britain: At the Edge of the World? 3000 BC to AD 1603* (London: BBC Worldwide, 2002), p. 20.

12. Sykes, *Saxons, Vikings, and Celts*, p. 137. Sykes incorrectly places Mount Sandel in County Antrim. It is actually in Coleraine, County Derry.

13. Bowen, *Britain and the Western Seaways*, p. 19; Barry A. Vann, *"Space of Time or Distance of Place": Presbyterian Diffusion in South-Western Scotland and Ulster, 1603–1690* (Berlin: VDM Verlag Dr. Muller, 2008), p. 17.

14. Sykes, *Saxons, Vikings, and Celts*, pp. 139–40.

15. Ibid., p. 136. It must be pointed out that Sykes is not willing to state that

this was a permanent settlement because 7000 BC predates the arrival of Neolithic peoples and their culture of plant domestication and, hence, agriculture to encourage permanent settlements.

16. Oliver S. Owen, *Natural Resource Conservation: An Ecological Approach*, 3rd ed. (New York: Macmillan, 1980), pp. 260–61.

17. Maire De Paor and Liam De Paor, *Early Christian Ireland* (New York: Praeger, 1958); Bowen, *Britain and the Western Seaways*, pp. 19–21.

18. Mary B. Dickinson, ed., *Wonders of the Ancient World: National Geographic Atlas of Archeology* (Washington DC: National Geographic Society, 1994).

19. Sykes, *Saxons, Vikings, and Celts*, p. 142.

20. Barry A. Vann, *In Search of Ulster-Scots Land: The Birth and Geotheological Imagings of a Transatlantic People, 1603–1703* (Columbia: University of South Carolina Press, 2008), pp. 155–58.

21. E. G. Ravenstein, "The Birth Places of the People and the Laws of Migration," *Geographical Magazine* 3 (1876): 173–77, 201–206, 229–33. See also his "Laws of Migration," *Journal of the Statistical Society* 48, no. 2 (1885): 167–235.

22. Wilbur Zelinsky, "The Hypothesis of Mobility Transition," *Geographical Review* 61, no. 2 (1971): 219–49; Huw Jones, "Evolution of Scottish Migration Patterns: A Social-Relations-of-Productions Approach," *Scottish Geographical Magazine* 102 (1986):151–64.

23. William A. Schwab, "The Origin of Cities," *The Sociology of Cities* (Englewood Cliffs, NJ: Prentice Hall, 1992) pp. 104–40.

24. Vann, *In Search of Ulster-Scots Land*.

25. Here, I am slightly modifying the terminology employed by E. G. Bowen in his *Britain and the Western Seaways*.

26. E. S. Beevy, "The Human Population," *Scientific American* 203, no. 3 (1960): 195–205; William A. Schwab, *The Sociology of Cities*, p. 107.

27. R. Braidwood, "From Cave to Village," cited in *Old World Archeology: Foundations of Civilizations*, ed. Charles Clifford Lamberg-Karlovsk (San Francisco: W. H. Freeman, 1972), pp. 67–70.

28. Dickinson, *Wonders of the Ancient World*, pp. 46, 62, 156, 178, 200, 276.

29. The oldest modern human remains found on the island of Great Britain date to about twelve thousand years ago. Found in a cave in Cheddar, England, the male remains have been dubbed "Cheddar Man."

30. Coriolis effect is caused by speed differentials among latitudinal points.

With twenty-four hours to rotate on its axis, the equator, with its much longer circumference, has to move faster than the comparatively shorter points north and south. For instance, the Arctic Circle has a substantially shorter circumference but has twenty-four hours to rotate. Fluid-moving bodies like ocean currents and winds are thus deflected to the right in the Northern Hemisphere and to the left in the Southern Hemisphere.

31. Sykes, *Seven Daughters of Eve*, p. 281.

32. Vann, *In Search of Ulster-Scots Land.*

33. Gordon Childe, "The Urban Revolution," *Town Planning Review* 21, no. 1 (1950): 3–17; J. T. Meyers, "The Origins of Agriculture: An Evaluation of Three Hypotheses," in *Prehistoric Agriculture*, ed. Stuart Struever (Garden City, NY: American Museum of Natural History, 1950), pp. 101–21.

34. Charles Burney, *The Ancient Near East* (Ithaca, NY: Cornell University Press, 1977), p. 11. Burney, along with Ellsworth Huntington, is one of a few Western scholars to have extensive field experience in Turkey. He was able to study firsthand the relics of ancient Anatolia. His attention to detail is impressive, but it does not make for pleasurable reading or geographic imaging.

CHAPTER 3. SETTLING THE ANCIENT ECUMENE

1. John R. Weeks, *Population: An Introduction to Issues and Concepts*, 8th ed. (Belmont, CA: Wadsworth, 2002), p. 8; James M. Rubenstein, *The Cultural Landscape: An Introduction to Human Geography*, 7th ed. (Upper Saddle River, NJ: Prentice Hall, 20020), p. 52.

2. Rubenstein, *Cultural Landscape*, p. 52.

3. Jerome D. Fellmann, Arthur Getis, and Judith Getis, *Human Geography: Landscapes of Human Activities*, 9th ed. (Boston: McGraw-Hill, 2007), p. 374.

4. C. Alan Joyce et al., eds., *World Almanac and Book of Facts, 2009* (Pleasantville, NY: World Almanac Education Group, 2009), pp. 546–55.

5. Ibid.

6. Ibid.

7. An analysis of variance (ANOVA) test was conducted on data for population density with respect to coastal, Great Lakes, and inland categories. The test showed that there is indeed a difference in population density by category (F [2, 46] = 8.88, p. < .001). By using a pair-wise model to compare means for the states'

population densities, independent samples *t*-tests showed that there was no difference between Great Lakes (M = 188.6) and coastal states (M = 337.3), although the small number of states classified as Great Lakes (n = 8) may well have affected the test, t (27) = 1.703, p > .05. On the other hand, a comparison between inland states (M = 46.4) and Great Lakes states (M = 188.6) was significantly different, t (27) = 4.41, p. < .001.

8. C. Allen Joyce et al., eds., *The World Almanac and Book of Facts 2009* (Pleasantville, NY: World Almanac Education Group, 2009), pp. 556–628.

9. Ibid., p. 717.

10. Ibid, p. 807.

11. Ibid, pp. 732–33.

12. Ibid, p. 717.

13. Paul L. Knox and Sallie A. Marston, *Human Geography: Places and Regions in Global Context*, 4th ed. (Upper Saddle River, NJ: Pearson Prentice Hall, 2007), p. 89.

14. Peter A. Morrison and Judith P. Wheeler, "Rural Renaissance in America? The Revival of Population Growth in Remote Areas" (Washington, DC: Population Reference Bureau, 1976).

15. An urban to rural migration flow clearly deviates from the gravity model of migration—a situation not recognized by some geographers. See, for example, Michael Kuby, John Harner, and Patricia Gober, *Human Geography in Action* (Danvers, MA: John Wiley and Sons, 2004), pp. 85–94; Robert E. Park, Ernest W. Burgess, and R. D. McKenzie, *The City* (Chicago: University of Chicago Press, 1925). Ferdinand Tönnies applied the German words *Gesellschaft* (society) and *Gemeinschaft* (community) to the study of the rural-urban dichotomy. In his conception of *Gemeinschaft*, a sense of community is associated with a close connection to the environment as well as to the informal structure of interpersonal relationships. The psychological bonds of community and attachment to space create something of a natural existence in which there is little need for formal laws. Urban places (*Gesellschaft*) feature detachments from both community and the environment, ushering in a need for formal laws and the apparatus of state to enforce common behaviors.

16. Rubenstein, *Cultural Landscape*, p. 44.

17. Joyce et al., *World Almanac 2009*, pp. 743, 748.

18. Ibid, p. 737.

19. Ibid, p. 797.

20. Ibid, p. 779.

21. Ibid, pp. 741, 824, 833.

22. Ibid, p. 745.

23. Ibid, p. 733.

24. Ibid, p. 744.

25. Fellmann, Getis, and Getis, *Landscapes of Human Activities*, p. 368.

26. Joyce et al., *World Almanac 2009*, p. 754.

27. Fellmann, Getis, and Getis, *Landscapes of Human Activities*, p. 368.

28. Joyce et al., *World Almanac 2009*, pp. 738, 791.

29. Ibid, pp. 566, 568.

30. Ibid, pp. 742, 767.

31. Ibid, pp. 767, 777.

32. Ibid, p. 594.

33. "Annual Climatological Data," National Climate Data Center, US Department of Commerce, cited in *World Almanac 2009*, eds. Joyce et al., p. 343.

34. Ibid, p. 628.

CHAPTER 4. GEOTHEOLOGY FROM
THE CRADLE OF CIVILIZATION

1. Lily Kong and others have provided well-written summaries of the work completed in the geographic study of religion. See Lily Kong, "Geography and Religion: Trends and Prospects," *Progress in Human Geography* 14, no. 3 (1990): 355–71; M. Pacione, "Relevance of Religion for a Relevant Human Geography," *Scottish Geographical Journal* 115 (1999): 117–31; Oliver Valins, "Identity, Space, and Boundaries: Ultra-Orthodox Judaism in Contemporary Britain" (PhD diss., University of Glasgow, UK, 1999). My own *In Search of Ulster-Scots Land: The Birth and Geotheological Imagings of a Transatlantic People, 1603–1703* (Columbia: University of South Carolina Press, 2007) also provides a discussion on how geographers study sacred space.

2. Carl O. Sauer, *Agricultural Origins and Dispersals* (New York: American Geographical Society, 1952), p. 1.

3. Carl O. Sauer, "Forward to Historical Geography," *Annals of the Association of American Geographers* 31 (1941): 1–24. On the links between Sauer and Bowen, see Colin Thomas, "Landscape with Figures: In the Steps of E. G. Bowen," *Cambria* 12 (1985): 24.

4. E. G. Bowen, "Le Pays de Galles," *Transactions and Papers of the Institute of British Geographers* 26 (1959): 1–23.

5. Vann, *In Search of Ulster-Scots Land*; Vann, *Puritan Islam: The Geoexpansion of the Muslim World* (Amherst, NY: Prometheus Books, 2011).

6. See John K. Wright, "Notes on Early American Geopiety," *Human Nature in Geography: Fourteen Papers, 1925–1965* (Cambridge: Harvard University Press, 1966); Yi-Fu Tuan, *Morality and Imagination: Paradoxes of Progress* (Madison: University of Wisconsin Press, 1989); Avihu Zakai, *Exile and Kingdom: History and Apocalypse in the Puritan Migration to America* (Cambridge: Cambridge University Press, 2002); Barry A. Vann, *In Search of Ulster-Scots Land: The Birth and Geotheological Imagings of a Transatlantic People, 1603–1703* (Columbia: University of South Carolina Press, 2007).

7. It is interesting that Chris Park does not offer any substantive discussion on the themes and topics advanced by Wright and even erroneously suggests that Tuan is the person associated with coining "geoteleology." See Park, *Sacred Worlds*, p. 19. It is perhaps important to make a distinction between what R. W. Stump calls religious geography and the geography of religions. Religious geography is concerned with understanding how religious thought worlds impact human actions that may have a direct influence on space. It therefore considers how the creation of imagined worlds impacts the landscape. On the other hand, the geography of religions identifies the spatial extent of religious groups (i.e., Christian denominations). Such an approach lends itself more to the creation of maps. See also Vann, *In Search of Ulster-Scots Land*.

8. John K. Wright, "Terrae Incognitae: The Place of Imagination in Geography," *Annals of the Association of American Geographers* 37 (1947): 1–15.

9. Ibid.

10. Vann, *In Search of Ulster-Scots Land*.

11. Genesis 1:1. All references from the Bible in this work, unless otherwise stated, are taken from *The Holy Bible: King James Version* (New York: American Bible Society, 2004).

12. Spencer Wells, *Deep Ancestry: Inside the Genographic Project* (Washington, DC: National Geographic Society, 2007), p. 106.

13. Genesis 6:7.

14. Daniel 7:14.

15. Revelation 11:15. Parts of this quote have been regularized to fit sentence-style capitalization. Italics are original to the text.

16. Surah 6:73. All references from the Qur'an in this work are taken from *The Qur'an Translation*, 14th ed. (Elmhurst, NY: Tahrike Tarsile Qur'an, 2003).

17. Surah 6:1–3.

18. Exodus 7:4; Leviticus 18:26; Numbers 21:2; Job 12:23; Jeremiah 49:36; Revelation 8:9.

19. Revelation 8:13.

20. Surah 5:120.

21. Surah 16:10–11.

22. Surah 16:49.

23. Surah 2:281.

24. Psalms 1:5–6.

25. Proverbs 29:26.

26. Surah 25:10–11.

27. Surah 9:109.

28. Revelation 11:18–19.

29. Surah 7:179.

30. Surah 15:43–45.

31. Surah 74:30–31.

32. Surah 74:26–29.

33. See, for instance, Surah 35:33–36; Surah 56:11–50. Here, in verse 36, Muhammad makes his promise of virgins to believers who make it to paradise.

34. Surah 18:39.

35. Revelation 20:11–15.

36. Muhammad was unclear as to why they were banished from the garden.

37. Leviticus 25:17–22.

38. Surah 9:72.

39. Surah 9:3.

40. Surah 9:113.

41. Lamentations 5:8.

42. Lamentations 5:21–22.

43. Psalms 78:12–15.

44. Psalms 78:4–48; Psalms 105:4–11.

45. Henry Campbell Black, *Black's Law Dictionary*, 6th ed. (Saint Paul, Minnesota: West Publishing, 1990), p. 33.

46. Surah 6:59.

47. Surah 14:19–20.

48. Surah 18:40–41.

49. Surah 10:26–27.

50. Surah 10:73.

51. Surah 2: 22.

52. Surah 3:198.

53. Surah 10:12.

54. Surah 10:14.

55. Surah 10:14.

56. Surah 10:13.

57. Surah 10:65.

58. Haggai 1:111.

59. Psalms 83:14–15.

60. Surah 13:11–13.

61. Surah 2:266.

62. The authorship of the book of Job is not known for certain, and there certainly are no historical events or known persons referenced in the book to which scholars can attached a date. For instance, Andrew Bruce Davidson, *The Book of Job* (Cambridge, UK: Cambridge University Press, 1889), p. 54, argues that the voice and range of topics used in Job are too different from those used in Exodus, so assuming that Moses was the author is not realistic.

63. From the New International Version. See *The Holy Bible: New International Version Containing the Old Testament and the New Testament* (Grand Rapids, MI: Zondervan, 1978).

64. Job 6:16; Job 37:10; Job 38:29.

65. Jeremiah was written around 600 BCE. See Jeremiah 18:14.

66. Isaiah 55:10.

67. Proverbs 26:1.

68. Stephen Reynolds (archivist), Bethsaida Excavations Project, housed at the International Studies and Programs, University of Nebraska–Omaha. See the website at http://www.unomaha.edu/bethsaida/default.htm (accessed August 23, 2010).

69. Ynetnews.com, "'Mini-Tornado' Sweeps through Western Galilee," http://www.ynetnews.com/articles/0,7340,L-3236346,00.html/ (accessed December 1, 2010).

70. Daniel 7:9.

CHAPTER 5. GEOKOLASIS: THE CASES OF DONORA, LAKE ERIE, AND THE DUST BOWL

1. See in particular People and Planet, "Green League 2011," http://people andplanet.org/climate-change/ (accessed September 27, 2009). Specifically, see the People and Planet Green League 2009 program in the United Kingdom, "Rewarding UK Universities for Excellent Environmental Performance," International Association of Universities, http://archive.www.iau-aiu.net/sd/sd_initiatives.html/ (accessed December 15, 2011).

2. Nancy McArdle, "Outward Bound: The Decentralization of Population and Employment," Joint Center for Housing Studies, Harvard University, July 1999, http://www.jchs.harvard.edu/publications/communitydevelopment/mcardle _w99-5.pdf/ (accessed October 24, 2010). John Ivanko and Lisa Kivirist, *Rural Renaissance: Renewing the Quest for the Good Life* (Gabriola Island, BC: New Society Publishers, 2009).

3. The issue of meeting the needs and wants of Asia's rising middle class while balancing environmental issues is further discussed in Asian Development Bank, *Rise of Asia's Middle Class* (Manila, Phil.: Asian Development Bank, 2010).

4. Paul L. Knox and Sallie A. Marston, *Human Geography: Places and Regions in Global Context*, 4th ed. (Upper Saddle River, NJ: Pearson Prentice Hall, 2007), p. 491.

5. Michael Bradshaw et al., *Contemporary World Regional Geography: Global Connections, Local Voices*, 3rd ed. (Boston: McGraw-Hill, 2009), p. 186.

6. Thomas Reardon, C. Peter Timmer, Christopher B. Barrett, Julio Berdequé, "The Rise of Supermarkets in Africa, Asia, and Latin America," *American Journal of Agricultural Economics* 85, no. 5 (2003): 1140–46.

7. Jean-Marie Grether, Nicole A. Mathys, and Jaime de Melo, "Global Manufacturing SO_2 Emissions: Does Trade Matter?" Kiel Institute, Hindenburgufer, Ger., 2009, http://www.wto.org/english/res_e/reser_e/gtdw_e/wkshop08_e/ mathys_e.pdf/ (accessed October 25, 2010).

8. Ibid.

9. Oliver S. Owen, *Natural Resource Conservation: An Ecological Approach* (New York: Macmillan, 1980), pp. 588–89.

10. Ibid. See also Howard R. Lewis, *With Every Breath You Take* (New York: Crown, 1965).

11. Owen, *Natural Resource Conservation*, p. 588.

12. Lewis, *Every Breath You Take*.

13. Michael West, "London's Killer Fogs Gone: Thanks to Coal-Burning Ban," *Wisconsin State Journal* (April 1978).

14. Lewis, *Every Breath You Take*; Owen, *Natural Resource Conservation*, pp. 587–88.

15. As a child growing up in Detroit, I made hundreds of personal observations of my family and neighbors burning refuse in allies. In winter months, I often joined groups of friends around incineration barrels to warm ourselves.

16. Robert H. Boyle, "Rebirth of a Dead Lake," *Sports Illustrated*, November 4, 1974.

17. Owen, *Natural Resource Conservation*, p. 177.

18. Sandra George, "Lake Erie," Environmental Protection Agency, Washington, DC, 2004, http://www.epa.gov/glnpo/solec/solec_2004/presentations/Lake_Erie_(George).pdf/ (retrieved October 1, 2009).

19. Owen, *Natural Resource Conservation*, pp. 174–75.

20. Ibid., pp. 175–76.

21. H. A. Regier and W. L. Hartman, "Lake Erie's Fish Community: 150 Years of Cultural Stress," *Science* 180 (1973): 1248–55.

22. Owen, *Natural Resource Conservation*, p. 177.

23. International Joint Commission: Canada and the United States, "Treaties and Agreements," http://www.ijc.org/rel/agree/quality.html (accessed October 1, 2009).

24. Sandra George, "Lake Erie."

25. Chris Evans, "Research Team Plies Lake Erie in Search of Problems, Answers," *Plain Dealer* (Cleveland, OH), September 20, 2009, http://www.cleveland.com/editorials/plaindealer/index.ssf?/base/opinion/1253349258110300.xml&coll=2 (accessed October 1, 2009).

26. Ibid.

27. Ibid.

28. James Donovan, *A Terrible Glory: Custer and the Battle of Little Bighorn* (Boston: Little, Brown, 2008), p. 20.

29. Francis Paul Prucha, "Andrew Jackson's Indian Policy: A Reassessment," *Journal of American History* 56, no. 3 (1969): 527–39. See also Donovan, *Terrible Glory*, p. 15.

30. Stalwart environmentalists like Oliver S. Owen not often write about

Lincoln's ulterior motives. Instead Owen sees the settlement of Indian Territory in purely ecological terms. See his *Natural Resource Conservation*, p. 88.

31. Paula McSpadden Love, "Clement Vann Rogers, 1839–1911," *The Chronicles of Oklahoma* 48 (1970): 389–99.

32. John Vosburgh, *Living with Your Land* (New York: Scribner, 1968).

33. William A. McGeveran Jr., ed., *World Almanac and Book of Facts 2003* (New York: World Almanac Education Group, 2003), p. 382.

34. Local anthropologist Jimmy Jon Atkinson, conversation with the author, May 1, 2000, Claremore, Oklahoma.

35. Owen, *Natural Resource Conservation*, pp. 88–89, claims that the intervals occur every thirty-five years, but in 1990, he teamed up with Daniel D. Chiras, a professor at the University of Colorado at Denver, and lowered the interval to twenty-two years. See their *Natural Resource Conservation: An Ecological Approach*, 5th ed. (New York: Macmillan, 1990), p. 83.

36. Don Eddy, "Up from the Dust," *Reader's Digest* 37 (1940): pp. 20–22.

37. Ibid. Owen, *Natural Resource Conservation*, pp. 91–92.

38. M. M. Leighton, "Geology of Soil Drifting on the Great Plain," *Scientific Monthly* 47 (1938): 22–33.

39. Will Rogers on the Dust Bowl, see http://archive.democrats.com/view .cfm?id=2595/ (accessed October 8, 2009).

40. Owen and Chiras, *Natural Resource Conservation*, pp. 86–87.

41. Ibid., p. 87.

42. David Goldfield et al., *The American Journey: A History of the United States* (Upper Saddle River, NJ: Pearson Prentice Hall, 2006), p. 528.

43. North Plains Groundwater Conservation District, "Ogallala Aquifer," http://www.npwd.org/new_page_2.htm/ (accessed October 10, 2009).

44. Donald Wilhite, "The Ogallala Aquifer and Carbon Dioxide: Are Policy Responses Applicable?" in Michael H. Glantz, *Societal Responses to Regional Climate Change: Forecasting by Analogy* (Boulder, CO: Westview Press, 1988).

CHAPTER 6. EROSION AND DISPOSSESSION IN THE CUMBERLAND GAP AREA, 1933–1939

1. Inspired by the lexicon of John K. Wright, *geokolasis* is derived from the Greek words for earth and punishment. See John K. Wright, "Notes on Early

American Geopiety," *Human Nature in Geography: Fourteen Papers, 1925–1965* (Cambridge, MA: Harvard University Press, 1966).

2. Donald Worster, *Dust Bowl: The Southern Plains in the 1930s* (New York: Oxford University Press, 2004). Worster's book won the prestigious Bancroft Prize.

3. John Steinbeck, *The Grapes of Wrath* (New York: Viking Press, 1939).

4. Michael J. McDonald and John Muldowny, *TVA and the Dispossessed* (Knoxville: University of Tennessee Press, 1982).

5. H. C. Darby, *Relations of History and Geography: Studies in England, France, and the United States* (Exeter, UK: University of Exeter Press, 2002). See also Chris Philo, "History, Geography, and the Still Greater Mystery of Historical Geography," in *Human Geography: Society, Space, and Social Science*, ed. Derek Gregory, Ron Martin, and Graham Smith (Minneapolis: University of Minnesota Press, 1994).

6. Paul Boyer et al., *The Enduring Vision: A History of the American People from 1865*, 5th ed. (Boston: Houghton Mifflin, 2004), p. 753.

7. Nelson Lichtenstein, Susan Strasser, and Roy Rosenzweig, *Who Built America? Working People and the Nation's Economy, Politics, Culture, and Society since 1877* (New York: Worth, 2000), p. 402.

8. Oliver S. Owen and Daniel D. Chiras, *Natural Resource Conservation: An Ecological Approach*, 5th ed. (New York: Macmillan, 1990), p. 7.

9. McDonald and Muldowny, *TVA and the Dispossessed*.

10. Donald Davidson, *I'll Take My Stand: The South and the Agrarian Tradition, 75th Anniversary Edition* (Baton Rouge: Louisiana State University Press, 2006). The book was originally published in 1930.

11. Les Williamson, telephone conversation with the author, February 18, 2010, Corbin, Kentucky.

12. McDonald and Muldowny, *TVA and the Dispossessed*.

13. Data from 2,699 participants were generated in a "Family Removal Questionnaire" by TVA in 1934. See also McDonald and Muldowny, *TVA and the Dispossessed*, p. 91.

14. John Rice Irwin's reflections on Norris Basin residents in the 1930s, in McDonald and Muldowny, *TVA and the Dispossessed*, p. 65. Irwin opened the Museum of Appalachia in 1969. It is located about five miles from Norris Dam.

15. Ibid., p. 207.

16. Rufus Voiles (1912–1995), conversation with the author, 1995, Oliver Springs, Tennessee.

17. McDonald and Muldowny, *TVA and the Dispossessed*.

18. William A. Schwab, *The Sociology of Cities* (Englewood Cliffs, NJ: Prentice Hall, 1992), p. 12; see also Harlan Barrows, *Geography as Human Ecology* (Indianapolis: Bobbs-Merrill Reprint Series in Geography, 1923).

19. Alan R. H. Baker, *Geography and History: Bridging the Divide* (Cambridge: Cambridge University Press, 2003).

20. Schwab, *Sociology of Cities*.

21. Ibid., p. 12; for a further look at the differences between the sociocultural and neoorthodox approaches to spatial perceptions see C. T. Jonassen, "Cultural Variables in the Ecology of an Ethnic Group," *American Sociological Review* 14, no. 5 (1949): 32–41; Walter Firey, *Land Use in Central Boston* (Cambridge, MA: Harvard University Press, 1947); Herbert J. Gans, *The Urban Villagers: Group and Class in the Life of Italian Americans* (New York: Free Press, 1962); and Amos Hawley, *Human Ecology: A Theory of Community Structure* (New York: Ronald Press, 1950).

22. Neil M. Maher, *Nature's New Deal: Civilian Conservation Corps and the Roots of the American Environmental Movement* (Oxford: Oxford University Press, 2007).

23. Sara M. Gregg, "Uncovering the Subsistence Economy in the Twentieth-Century South: Blue Ridge Mountain Farms," *Agricultural History Society* 78, no. 4 (2004): 417–37.

24. The concept of geography as human ecology has been further explored by L. F. Schnore, "Geography and Human Ecology," *Economic Geography* 37, no. 3 (July 1961): 207–17; Philip W. Porter, "Geography as Human Ecology: A Decade of Progress in a Quarter Century," 22, no. 1 (1978): 15–39. Karl. S. Zimmerer, "Human Geography and the New Ecology: The Prospect and Promise of Integration," *Annals of the Association of American Geographers* 84, no.1 (March 1994): 108–25.

25. Colleges included Lincoln Memorial University and the University of the Cumberlands.

26. Franklin D. Roosevelt, *On Our Way* (New York: John Day, 1934) quoted in McDonald and Muldowny, *TVA and the Dispossessed*, p. 11.

27. While there are a number of well-written books on the TVA, Preston J. Hubbard devotes a great deal of attention to its inception. See his *Origins of the TVA: The Muscle Shoals Controversy, 1920–1932* (New York: W. W. Norton, 1961).

28. Rufus Voiles, conversation with the author.

29. Roosevelt, *On Our Way*.

30. Roy Talbert, "Arthur E. Morgan's Ethical Code for the Tennessee Valley Authority," *Publications of the East Tennessee Historical Society* 40 (1968): 119–27. See also his master's thesis titled "Human Engineer: A. E. Morgan and the Launching of the Tennessee Valley Authority" (master's thesis, Vanderbilt University, 1967).

31. Despite his prominence in American domestic policy, few books have been written on David E. Lilienthal. While lacking appeal for a general readership, Steven M. Neuse has written the most extensive academic work to date on Lilienthal. See his *David E. Lilienthal: The Journey of an American Liberal* (Knoxville: University of Tennessee Press, 1996).

32. Wayne Flynt, *Dixie's Forgotten People: The South's Poor Whites* (Bloomington: Indiana University Press, 2004), p. 1.

33. Verna Mae Slone, *What My Heart Wants to Tell* (Boston: G. K. Hall, 1979). The book was republished by the University Press of Kentucky in 1988.

34. Ibid. Barry A. Vann, *Rediscovering the South's Celtic Heritage* (Johnson City, TN: n.p., 2004); Wayne J. Flynt, *Dixie's Forgotten People* (Bloomington: Indiana University Press, 1980); Harry M. Caudill, *Night Comes to the Cumberlands: A Biography of a Depressed Area* (Boston: Little, Brown, 1963).

35. Odette Keun, *A Foreigner Looks at the TVA* (New York: Longmans, Green, 1937), p. 30.

36. Sloane, *What My Heart Wants to Tell*; Ronald D. Eller, *Miners, Millhands, and Mountaineers: Industrialization of the Appalachian South, 1880–1930* (Knoxville: University of Tennessee Press, 1982); Flynt, *Dixie's Forgotten People*; Caudill, *Night Comes to the Cumberlands*; Henry D. Shapiro, *Appalachia on Our Mind: The Southern Mountains and Mountaineers in the American Consciousness, 1870–1920* (Chapel Hill: University of North Carolina Press, 1978).

37. Nathaniel Hirsch, "An Experimental Study of the East Kentucky Mountaineers," *Genetic Psychology Monographs* 3 (March 1928): 183–244; Vann, *Rediscovering the South's Celtic Heritage*.

38. McDonald and Muldowny, *TVA and the Dispossessed*, p. 7.

39. Les Williamson, conversation with the author.

40. Charles Lane, who lives in Johnson County, Kentucky, has ten siblings. Six of them left southern Appalachia in the 1950s and 1960s to work in automobile plants in the Detroit area. Conversation with the author, December 12, 2009, Speedwell, Tennessee.

41. David Goldfield et al., *American Journey: A History of the United States* (Upper Saddle River, NJ: Pearson Prentice Hall, 2006), p. 520.

42. Linda Keck (née Richardson) is the principal of Cumberland Gap High School and a resident of Claiborne County, Tennessee. Conversation with the author, February 19, 2010, Speedwell, Tennessee.

43. Vernedith Voiles (1919–1999), resident of Anderson County, Tennessee, conversations with the author, July 14–15, 1997, Oliver Springs, Tennessee. Characterizing these farms as either subsistence or commercial enterprises is difficult to do because of the informal nature of the regional culture. Profits and other forms of income often went unreported to the government, so financial data on the impact of Norris Dam on the agricultural economy is not a valid estimate of real income.

44. Mary Ruth Isaacs, e-mailed survey response, February 20, 2010, Williamsburg, Kentucky.

45. Les Williamson, conversation with the author.

46. Terri Keeton, e-mailed survey response, February 27, 2010, Oneida, Tennessee.

47. Dorothy Jones, conversation with the author, January 10, 2010, Oliver Springs, Tennessee.

48. Jimmie Edwards, conversation with the author, February 19, 2010, Union County, Tennessee.

49. Rosemary Weddington, e-mailed survey response, February 20, 2010, Frankfort, Kentucky.

50. Ibid.

51. Oline Carmical, e-maied survey response, February 17, 2010, Williamsburg, Kentucky.

52. Sue Wake, e-mailed survey response, February 17, 2010, Williamsburg, Kentucky.

53. Diana Baker, e-mailed survey response, February 17, 2010, Williamsburg, Kentucky.

54. Harold Hubbard, e-mailed survey response, February 18, 2010, Williamsburg, Kentucky.

55. Gregg, "Uncovering the Subsistence Economy."

56. Nancy Maiden, owner-broker of Realty Network, Harrogate, Tennessee, e-mailed survey response, December 2, 2009.

57. Ibid.

58. Appalachian Regional Commission, "Per Capita Income Rates in Appalachia, 2008," http://www.arc.gov/research/MapsofAppalachia.asp?MAP_ID=56/ (accessed November 28, 2009).

59. Ayers Auction and Real Estate, 2009, http://www.ayersrealtypage .com/propview.php?show_thumbs=1/ (accessed December 3, 2009). The property is located in Powell Valley.

60. Karen Edwards, interview with the author, November 25, 2009, Speedwell, Tennessee.

CHAPTER 7. LIFE AND DEATH IN HURRICANE ALLEY

1. C. Alan Joyce et al., eds., *World Almanac and Book of Facts 2009* (Pleasantville, NY: World Almanac Education Group, 2009), p. 353.

2. The Saffir-Simpson Hurricane Scale is based on wind velocity and storm surge.

3. National Hurricane Center, "NHC Data Archive," NOAA, http://www .nhc.noaa.gov/HAW2/english/history.shtml#katrina/ (accessed June 16, 2010).

4. Ibid. According to the National Hurricane Center, one thousand were killed in Louisiana and two hundred died in Mississippi.

5. Richard D. Knabb, Jamie R. Rhome, and Daniel P. Brown, "Tropical Cyclone Report: Hurricane Katrina," The National Hurricane Center, December 20, 2005. See http://www.nhc.noaa.gov/HAW2/english/history.shtml#katrina/ (accessed June 16, 2010).

6. C. Alan Joyce et al., *World Almanac 2008*, p. 303.

7. Steve Bird, "Cyclone Syd Destroys Homes in Bangladesh," *Times of London: Sunday Edition*, November 15, 2007. See "The Bangladesh Cyclone of 1991," US Aid, http://pdf.usaid.gov/pdf_docs/PNADG744.pdf/ (accessed December 15, 2011).

8. Joyce et al., *World Almanac 2008*, p. 303.

9. MCEER Report, "Myanmar (Burma) Cyclone Nargis Disaster 2008," Research Foundation of the State of New York, http://mceer.buffalo.edu/ infoservice/disasters/burma_cyclone.asp/ (accessed August 26, 2010).

10. In Asia, typhoon classification is based on a three-tier system: typhoon (118–149 kmh); severe typhoon (150–184 kmh); super typhoon (185 or above). A wind speed of 118 kmh is converted and rounded to 74 mph. See the system employed by the Hong Kong Observatory at http://www.weather.gov.hk/ aviat/outreach/product/20th/TCclass.htm/ (accessed June 17, 2010).

11. Alan Strahler and Arthur Strahler, *Introducing Physical Geography*, 3rd

ed. (Hoboken, NJ: John Wiley and Sons, 2003), p. 201.

12. Of the sixty-five major hurricanes to make landfall in the United States between 1900 and 2000, some of them struck more than one state. Consequently, individual states along the Atlantic and Gulf were hit ninety-five times. Specifically, states on the Gulf coast were hit fifty-five times, and states along the Atlantic were struck forty times. Data were analyzed from a NOAA report by Jerry D. Jarrell, Max Mayfield, and Edward N. Rappaport, "The Deadliest, Costliest, and Most Intense United States Hurricanes from 1900 to 2000 (and Other Frequently Requested Hurricane Facts)," NOAA, Technical Memorandum, 2001, NWS TPC-1. See Table 10 of the Report. The question as to whether states located on the Gulf are more vulnerable to hurricanes than their Atlantic counterparts is answered by z-test for Two Proportions Test: $z = 2.03$; one tail test confidence level is 97.9%. See http://www.aoml.noaa.gov/hrd/Landsea/deadly/index.html/ (accessed August 4, 2010).

13. Coriolis effect is seen when fluid moving objects such as water or wind currents are deflected to the right in the Northern Hemisphere and to the left in the southern half of the globe. This is caused by the different speeds that the earth's surface moves in rotation relative to latitudinal location. For instance, the earth has a much larger circumference at the equator than at the tropics. The circumference is far shorter at the Arctic and Antarctic circles, yet each latitudinal point has twenty-four hours to move along its rotational path. This means that any given point along the equator is moving in a counterclockwise direction (west to east) at about 1,040 mph, while at 60°north, the earth's speed is only 500 mph. Ocean currents along the tropics are moving much faster than 500 mph, so as these currents move north, they pass over earth that is moving slower to the east while the speed of the current is still much higher. The result is that the current seems to be deflected to the right even though it is moving straight. See Strahler and Strahler, *Introducing Physical Geography*, pp. 158–59.

14. Jarrell, Mayfield, and Rappaport. "Deadliest, Costliest and Most Intense United States Hurricanes."

15. Ibid.

16. Sebastian Yunger, *The Perfect Storm: A True Story of Men against the Sea* (New York: HarperCollins, 1997).

17. Strahler and Strahler, *Introducing Physical Geography*, p. 53. *Encyclopedia of Earth: Content, Credibility, Community*, April 15, 2010, http://www.eoearth.org/article/Energy_balance_of_Earth/ (accessed December 15, 2011).

18. Michael Pidwirny, "Energy Balance of Earth," *Encyclopedia of Earth: Content, Credibility, Community*, April 15, 2010, http://www.eoearth.org/article/Energy_balance_of_Earth (accessed December 15, 2011).

19. A good discussion of this debate in the press is provided by Jeffrey Kluger, "Is Global Warming Fueling Katrina?" *Time*, August 29, 2005, http://www.time.com/time/nation/article/0,8599,1099102,00.html/ (accessed July 14, 2010).

20. Julie Snider, "Global Warming Activists Turn Storms into Spin," *USA Today*, September 25, 2005. See the article at http://www.usatoday.com/news/opinion/editorials/2005-09-25-our-view_x.htm/ (accessed July 15, 2010).

21. Ibid.

22. Kluger, "Is Global Warming Fueling Katrina?"

23. Ibid.

24. Kenneth Trenberth, "Climate: Uncertainty in Hurricanes and Global Warming," *Science* 308, no. 5729 (2005): 1753–54.

25. Richard B. Stothers, "The Great Tambora Eruption in 1815 and Its Aftermath," *Science* 224, no. 4654 (1984): 1191–98; Clive Oppenheimer, "Climatic, Environmental and Human Consequences of the Largest Known Historic Eruption: Tambora Volcano (Indonesia) 1815," *Progress in Physical Geography* 27, no. 2 (2003): 230–59.

26. C. Edward Skeen, "The Year without a Summer: A Historical View," *Journal of the Early Republic* 1, no. 1 (1981): 51–67.

27. Paul Leicester Ford, ed., "Jefferson to Gallatin," September 8, 1816, *The Writings of Thomas Jefferson* vol 10 of 10 (New York, 1892–1899), pp. 64–65, in C. Edward Skeen, "Year without a Summer."

28. Willie Soon and Steven H. Yaskell, "A Year without a Summer: A Weak Solar Maximum, a Major Volcanic Eruption, and Possibly Even the Wobbling of the Sun Conspired to Make the Summer of 1816 One of the Most Miserable Ever Recorded," *Mercury* (May–June 2003): 13–22.

29. Andrés Poey, "A Chronological Table, Comprising 400 Cyclonic Hurricanes Which Have Occurred in the West Indies in the North Atlantic within 362 Years, from 1493 to 1855," *Journal of the Royal Geographical Society* 25 (1855): 291–328.

30. Michael Chenoweth, "A Reassessment of Historical Atlantic Basin Tropical Cyclone Activity, 1700–1855," NOAA (2006), http://www.aoml.noaa.gov/hrd/hurdat/Chenoweth/chenoweth06.pdf/ (accessed July 22, 2010).

31. Snider, "Global Warming Activists Turn Storms into Spin."

32. Zones of energy deficits form when long-wave radiation leaving the atmosphere exceeds the amount of solar energy (short-wave radiation) that enters the biosphere. This is called "net radiation." The earth's energy balance is also affected by albedo (the reflectivity of surfaces) and greenhouse gases such as humidity and carbon dioxide.

33. Erik Larson, *Isaac's Storm: A Man, a Time, and the Deadliest Hurricane in History* (New York: Vintage Books, 1999), p. 65.

34. Ibid., p. 66.

35. Ibid., p. 9.

36. Ibid., p. 13.

37. Ibid., pp. 13, 144.

38. Ibid., p. 20.

39. Ibid., p. 9.

40. Walter W. Davis, cited in ibid., pp. 148–49.

41. Ibid., p. 149.

42. Isaac M. Cline, "Special Report on the Galveston Hurricane of September 8, 1900," NOAA History Files, http://www.history.noaa.gov/stories _tales/cline2.html/ (accessed August 2, 2010).

43. Ibid.

44. Ibid.

45. Ibid.

46. Larson, *Isaac's Storm*.

47. Jarrell, Mayfield, and Rappaport. "Deadliest, Costliest and Most Intense United States Hurricanes."

CHAPTER 8. LIFE AND DEATH IN AMERICA'S TORNADO ALLEYS

1. Tom Henderson, "Tornado Timeline: Updated," Chattanooga News Channel 9 May 4, 2011, http://www.newschannel9.com/articles/timeline-1000 891-tornado-55am.html/ (accessed May 30, 2011).

2. Greg Bluestein and Melissa R. Nelson, "Survivors Picking up Pieces from Deadly Twisters," Associated Press, April 29, 2011, http://news.yahoo.com/s/ap/ ap_on_re_us/us_severe_weather/ (accessed April 29, 2011). Associated Press, "South Struggles to Rebound from Deadliest Tornado Outbreak since Great

Depression," Fox News, April 29, 2011, http://www.foxnews.com/us/2011/04/29/survivors-picking-pieces-deadly-twisters-kill-27/ (accessed April 29, 2011). Weather on MSNBC.com, "Twister Outbreak Is Second Deadliest in US History," April 30, 2011, http://www.msnbc.msn.com/id/42834400/ns/weather/?GT1 =43001/ (accessed April 30, 2011).

3. Weather on MSNBC.com, "Twister Outbreak."

4. National Oceanic and Atmospheric Administration (NOAA), "Storm Survey, May 22, 2011," National Weather Service, Springfield, Missouri, May 22, 2011, http://www.crh.noaa.gov/sgf/?n=event_2011may22_survey/ (accessed May 26, 2011). See the revised page at http://www.crh.noaa.gov/sgf/?n=event _2011may22_summary/ (accessed June 1, 2011).

5. National Oceanic and Atmospheric Administration (NOAA), "April 2011 Tornado Information," http://www.noaanews.noaa.gov/april_2011 _tornado_information.html (accessed May 1, 2011).

6. Dorothy and Toto are characters in the 1939 motion picture *The Wizard of Oz*, produced by Victor Fleming (Burbank, CA: Warner Brothers, 1939). Bill and Jo are characters in the 1996 film *Twister*, directed by Jan de Bont (Burbank, CA: Warner Brothers, 1996).

7. Storm Prediction Center, "The 25 Deadliest U.S. Tornadoes," NOAA, http://www.spc.noaa.gov/faq/tornado/killers.html/ (accessed December 15, 2011).

8. Harold E. Brooks and Charles A. Doswell III, "Normalized Damage from Major Tornadoes in the United States: 1890–1999," *Weather and Forecasting* 16, no. 1 (2001): 168–76.

9. Wallace Akin, *Forgotten Storm: The Great Tri-State Tornado of 1825* (Guilford, CT: Lyons Press, 2002).

10. National Oceanic and Atmospheric Administration (NOAA), "History of Tornado Forecasting," http://celebrating200years.noaa.gov/magazine/tornado _forecasting/welcome.html#knowledge/ (accessed November 22, 2010).

11. Ibid.

12. Ibid.

13. C. Allen Joyce et al., eds., *World Almanac and Book of Facts 2009* (Pleasantville, NY: World Almanac Books), p. 352.

14. Storm Prediction Center, "25 Deadliest Tornadoes." http://www.spc .noaa.gov/faq/tornado/killers.html/ (accessed December 15, 2011).

15. NewsOK.com, "May 3rd," OPUBCO Communications Group, http://newsok.com/may3/ (accessed November 3, 2010).

16. Thomas P. Grazulis, *Tornado: Nature's Ultimate Windstorm* (Norman: University of Oklahoma Press, 2001), p. 8.

17. NOAA, "History of Tornado Forecasting."

18. Increase Mather, *Remarkable Providences: An Essay for the Recording of Illustrious Providences* (Boston,1674). Latest publications by George Lincoln Burr, ed., *Narratives of the Witchcraft Cases 1648–1706* (New York: Charles Scribner's Sons, 1914), pp. 3–38.

19. Tornado Chaser, "History of Tornadoes," http://www.tornadochaser .net/history.html/ (accessed November 23, 2010).

20. James R. McDonald, "T. Theodore Fujita: His Contribution to Tornado Knowledge through Damage Documentation and the Fujita Scale," *Bulletin of the American Meteorological Society* 82, no. 1 (2001): 63–72.

21. National Oceanic and Atmospheric Administration (NOAA), "Enhanced Fujita Scale Overview," http://www.erh.noaa.gov/rah/news/content/ Enhanced.Fujita.Scale.Overview.html/ (accessed December 11, 2010).

22. National Geographic Society, "Forces of Nature: Tornadoes," http:// www.nationalgeographic.com/forcesofnature/interactive/index.html/ (accessed November 19, 2010). Not all tornadoes are produced by supercells; for a discussion them, see Roger M. Wakimoto and James W. Wilson, "Non-Supercell Tornadoes," *Journal of Atmospheric Sciences* 117 (1989): 1113–40.

23. National Geographic Society, "Forces of Nature: Tornadoes."

24. This comment is based on a list of deadly tornadoes published in Joyce et al., *World Almanac 2009*, p. 352.

25. Ibid.

26. Ibid.

27. Akin, *Forgotten Storm*, p. 8.

28. Brian Williams, *Nightly News*, NBC, April 28, 2011, http://video .tvguide.com/NBC+Nightly+News+with+Brian+Williams/April+Storms +Break+Weather+Records/7195883?autoplay=true&partnerid=OVG/ (accessed April 30, 2011).

29. National Geographic Society, "Tornadoes," http://www.national geographic.com/forcesofnature/interactive/index.html/ (accessed November 21, 2010).

30. Akin, *Forgotten Storm*, p. 25.

31. Peter Felknor, *The Tri-State Tornado: The Story of America's Greatest Tornado Disaster* (New York: iUniverse, 2004), p. 4.

32. Akin, *Forgotten Storm*, pp. 26–27.

33. Felknor, *Tri-State Tornado*, p. 5.

34. Ibid., p. 7.

35. Ibid., p. 9.

36. Akin, *Forgotten Storm*.

37. Ibid.

38. Ibid., pp. 112–14.

39. Dennis Cauchon and Traci Watson, "Deadly Tornadoes Hit Several States," *USA Today*, January 1, 2003, http://www.usatoday.com/weather/tornado/2002-11-11-deadly-storm_x.htm/ (accessed November 28, 2010).

40. Ibid.

41. National Weather Service Forecast Office, "Cumberland Plateau Tornado Outbreak of November 11, 2002," http://www.srh.noaa.gov/mrx/svrevnts/nov10tornadoes/nov10tornado.php/ (accessed November 28, 2010).

42. CNN, "Residents Clean up after Killer Storms," November 11, 2002, http://articles.cnn.com/2002-11-11/weather/severe.storms_1_tornado-mossy-grove-van-wert-county?_s=PM:WEATHER/ (accessed November 29, 2010).

43. Steven Williamson and Sarah Vann, conversation with the author, December 22, 2010, Mossy Grove, Tennessee.

44. Ibid.

45. Cauchon and Watson, "Deadly Tornadoes."

CHAPTER 9. LIFE AND DEATH IN QUAKE ZONES

1. Alan Strahler and Arthur Strahler, *Introducing Physical Geography*, 3rd ed. (New York: John Wiley and Sons, 2003), p. 398.

2. National Geographic News, "The Deadliest Tsunami in History?" *National Geographic*, January 7, 2005, http://news.nationalgeographic.com/news/2004/12/1227_041226_tsunami.html/ (accessed December 6, 2010).

3. CBSNews.com Staff, "Japan Upgrades Killer Quake's Magnitude," March 13, 2011, http://www.cbsnews.com/8301-503543_162-20042549-503543.html/ (accessed April 15, 2011).

4. Justin McCurry, "Japan Quake Death Toll Passes 18,000," *Guardian*, March 23, 2011, http://www.guardian.co.uk/world/2011/mar/21/japan-earthquake-death-toll-18000/ (accessed April 1, 2011).

5. Fox News, "Tsunami Slams Northern Japan after Massive 8.9 Magnitude Earthquake Strikes off Coast," March 11, 2011, http://www.foxnews.com/world/2011/03/11/massive-7-magnitude-earthquake-strikes-japan/ (accessed March 11, 2011).

6. Malcolm Foster, "60 Killed in Major Tsunami after 8.9 Japan Quake," Associated Press, March 11, 2011, http://news.yahoo.com/s/ap/ap_on_re_as/as_japan_earthquake/ (accessed March 11, 2011).

7. Earthquake Hazards Program, USGS, "Historic Earthquakes: New Madrid Earthquakes 1811–1812," (paragraph 1), http://earthquake.usgs.gov/earthquakes/states/events/1811-1812.php/ (accessed December 7, 2010).

8. Peter M. Shearer. *Introduction to Seismology* (Cambridge, UK: Cambridge University Press, 1999).

9. H. C. Darby, *Relations of History and Geography: Studies in England, France, and the United States* (Exeter, UK: University of Exeter Press, 2002). See also Chris Philo, "History, Geography, and the Still Greater Mystery of Historical Geography," in *Human Geography: Society, Space, and Social Science*, ed. Derek Gregory, Ron Martin, and Graham Smith (Minneapolis: University of Minnesota Press, 1994).

10. Eugene Schweig, "Scientists Update New Madrid Earthquake Forecasts," USGS Newsroom, January 13, 2003, http://www.usgs.gov/newsroom/article.asp?ID=215 (accessed December 8, 2010).

11. For a more in-depth examination of the Center for Cave and Karst Studies, see Hoffman Environmental Research Institute, Western Kentucky University, http://caveandkarst.wku.edu/.

12. Shearer, *Introduction to Seismology*, p. 2.

13. Robert Mallet, *Great Neapolitan Earthquake of 1857: The First Principles of Observational Seismology* (London: Chapman and Hall, 1862).

14. Science for a Successful Ireland, "Dalkey Memorial to Seismology Pioneer," Science.ie, http://www.science.ie/science-news/seismology-pioneer-robert-mallet.html/ (accessed December 11, 2010).

15. Scientific Itineraries in Tuscany, "Filippo Cecchi," http://brunelleschi.imss.fi.it/itineraries/biography/FilippoCecchi.html/ (accessed December 12, 2010).

16. Shearer, *Introduction to Seismology*, p. 2.

17. F. Du Bois, "The Cecchi Seismograph: A New Model Constructed for the Observatory of Manila," *Transactions of the Seismological Society of Japan* 5–8 (1885): 90–94.

18. Agustin Udias and William Stauder, "The Jesuit Contribution to Seismology," *International Geophysics* 81, no. 1 (2002): 19–27.

19. Shearer, *Introduction to Seismology*, p. 10.

20. Committee on Seismology, *Seismology: Responsibilities and Requirements of a Growing Science* (Washington, DC: Division of Earth Sciences National Academy of Sciences, 1969), p. 6.

21. USGS Earthquake Hazards Program, "The Richter Magnitude Scale," http://earthquake.usgs.gov/learn/topics/richter.php/ (accessed December 14, 2010).

22. Ibid.

23. USGS, "Modified Mercalli Intensity Scale," http://earthquake.usgs.gov/learn/topics/mercalli.php/ (accessed December 22, 2010).

24. Schweig, "Scientists Update New Madrid Earthquake Forecasts."

25. Otto W. Nuttli, "The Mississippi Valley Earthquakes of 1811 and 1812: Intensities, Ground Motion and Magnitudes," *Bulletin of the Seismological Society of America* 63, no. 1 (1973): 227–48.

26. Jay Feldman, *When the Mississippi Ran Backwards: Empire, Intrigue, Murder, and the New Madrid Earthquakes* (New York: Free Press, 2005), p. 145.

27. Earthquake Hazards Program, "Historic Earthquakes," USGS, http://earthquake.usgs.gov/earthquakes/states/events/1811-1812.php/ (accessed December 23, 2010).

CHAPTER 10. CHALLENGES OF CLIMATE CHANGE

1. Mike Hulme, "The Conquering of Climate: Discourses of Fear and Their Dissolution," *Geographical Journal* 174, no. 1 (March, 2002): 5–16.

2. Intergovernmental Panel on Climate Change, "History," United Nations, http://www.ipcc.ch/organization/organization_history.shtml/ (accessed January 10, 2011).

3. *The Lion King*, directed by Roger Allers and Rob Minkoff (Burbank, CA: Walt Disney Pictures, 1994).

4. Aldo Leopold, *A Sand County Almanac: And Sketches Here and There* (New York: Oxford University Press, 1949).

5. Ibid., p. 204. For an excellent review and critical assessment of *A Sand County Almanac*, see J. Baird Callicott, ed., *Companion to A Sand County*

Almanac: Interpretive and Critical Essays (Madison: University of Wisconsin Press, 1987).

6. Tim Ingold, *The Perception of the Environment: Essays on Livelihood, Dwelling and Skill* (New York: Routledge, 2000), pp. 112–13.

7. Hulme, "Conquering of Climate."

8. Émile Durkheim, Carol Cosman, and Mark Sydney Cladis, *The Elementary Forms of Religious Life* (Oxford: Oxford University Press, 2002).

9. Guillermo Gonzales and Jay Wesley Richards, *The Privileged Planet: How Our Place in the Cosmos Is Designed for Discovery* (Washington, DC: Regnery Press, 2004).

10. Jo Robinson, *Pasture Perfect: The Far-Reaching Benefits of Choosing Meat, Eggs, and Dairy Products from Grass-Fed Animals* (Vashon, WA: Vashon Island Press, 2004), p. 51.

11. John Raines, ed., *Marx on Religion* (Philadelphia, PA: Temple University Press, 2002).

12. Oliver S. Owen, *Natural Resource Conservation: An Ecological Approach*, 3rd ed. (New York: Macmillan, 1980).

13. Paul Kirshen, Kelly Knee, and Mathias Ruth, "Climate Change and Coastal Flooding in Metro Boston: Impacts and Adaptation Strategies," *Climate Change* 90, no. 4 (2008): 453–73.

14. Brian K. Sullivan, "NYC Snow Storm Sets Record, Stops Flights, Cuts Power," *Bloomberg Business Week*, February 27, 2010, http://www.businessweek .com/news/2010-02-27/nyc-snow-storm-sets-record-stops-flights-cuts-power update1-.html/ (accessed January 2, 2011).

15. Daniel Trotta, "New York Hard Hit as Winter Storm Slams Northeast," *Reuters: Edition US*, December 27, 2010, http://www.reuters.com/article/ idUSTRE6BP1EW20101227/ (accessed January 4, 2011).

16. Ibid.

17. Ibid.

18. Daniel Trotta, "Record Snow Again Buries Northeast," *Reuters*, January 27, 2011, http://www.reuters.com/article/2011/01/27/us-usa-weather-business -idUSTRE70Q6M120110127/ (accessed January 29, 2011).

19. "New Monthly Snowfall Record Set in State," WFSB Channel 3, *Eyewitness News*, January 27, 2011, Hartford, Connecticut, http://www.wfsb.com/ news/26637950/detail.html/ (accessed January 29, 2011).

20. Trotta, "Record Snow Again Buries Northeast."

21. Fox News.com, "Northeast Braces for Powerful Snowstorm That Paralyzed the South," http://www.foxnews.com/weather/2011/01/11/northeast-braces-powerful-snowstorm-paralyzed-south/ (accessed January 13, 2011).

22. MSNBC, "2 Blizzards in a Week? Central U.S. Sees 'Very Unusual' Repeat," Weather, MSNBC, February 9, 2011, http://www.msnbc.msn.com/id/41486321/ns/weather/?gt1=43001/ (accessed February 9, 2011).

23. Lydia Saad, "Water Issues Worry Americans Most, Global Warming Least," Gallup, March 28, 2011, http://www.gallup.com/poll/146810/Water-Issues-Worry-Americans-Global-Warming-Least.aspx/ (accessed April 5, 2011).

24. William D. Nordhaus and Zili Yang, "A Regional Dynamic General-Equilibrium Model of Alternative Climate Change Strategies," *American Economic Review* 86, no. 4 (1996): 741–65.

25. Richard B. Stothers, "The Great Tambora Eruption in 1815 and Its Aftermath," *Science* 224 (1984), pp. 1191–98.

26. Alan Strahler and Arthur Strahler, *Introducing Physical Geography*, 3rd ed. (New York John Wiley and Sons, 2003), p. 109.

27. Ibid., p. 108.

28. While they do not make an argument for down-cycling CO_2 and H_2O as a factor in climate change, Charles A. Perry and Kenneth J. Hsu argue that solar activity is the major cause of climate change. This is a notion shared by this author (see chapter 8). Their work is cited here in regard to the existence of eight ice ages over the past 720,000 years. See their "Geophysical, Archaeological, and Historical Evidence Support a Solar-Output Model for Climate Change," *Proceedings of the National Academy of Sciences of the United States of America* 97, no. 23 (2000): 12433–38.

29. Ralph Buchsbaum, *Animals without Backbones: An Introduction to the Invertebrates* (Chicago: University of Chicago Press, 1948).

30. Mark H. Scheihing and Hermann W. Pfefferkorn, "Taphonomy of Land Plants in the Orinoco Delta: A Model for the Incorporation of Plant Parts in Clastic Sediments of Late Carboniferous Age of Euramerica," *Review of Palaeobotany and Palynology* 41, nos. 3–4 (1984): 205–40.

31. Strahler and Strahler, *Introducing Physical Geography*.

32. Goddard Institute for Space Studies, "2009: Second Warmest Year on Record; End of Warmest Decade," NASA, January 21, 2010, http://www.giss.nasa.gov/research/news/20100121/ (accessed January 6, 2011).

33. Ibid.

34. Sullivan, "NYC Snow Storm Sets Record." For the 1940 to 1980 time period, see Owen, *Natural Resource Conservation*, p. 579.

35. Mallory Simon, "This Just in: NFL Postpones Eagles-Vikings Game until Tuesday," CNN, December 26, 2010, http://news.blogs.cnn.com/2010/12/26/nfl-postpones-eagles-vikings-game-until-tuesday/ (accessed January 7, 2011).

36. Daniel Carty, "Metrodome Roof Collapse Caught on Tape," CBS Sports, December 13, 2010, http://www.cbsnews.com/8301-31751_162-20025435-10391697.html/ (accessed January 10, 2011).

37. Weather Forecast Office of the National Weather Service, "Jacksonville, FL," http://www.srh.noaa.gov/jax/?n=climate/ (accessed January 12, 2011).

38. Lloyd D. Keigwin, "The Little Ice Age and Medieval Warm Period in the Saragossa Sea," *Science* 274, no. 5292 (1996):1503–1508.

39. Perry and Hsu, "Geophysical, Archaeological, and Historical Evidence Support a Solar-Output Model for Climate Change."

40. Earl D. Radmacher, Ronald B. Allen, and H. Wayne House are theology professors at Western Conservative Baptist Seminary, Dallas Theological Seminary, and Michigan Theological Seminary, respectively. They are also Bible editors for Thomas Nelson, Inc. Their analysis of the book of Job is found on page 769 of the NKLV Study Bible, 2nd. ed. (Nashville: Thomas Nelson, 2007).

41. Perry and Hsu, "Geophysical, Archeological, and Historical Evidence Support a Solar-Output Model for Climate Change," p. 12435.

42. Kenneth J. Hsu, *Climate and Peoples: A Theory of History* (Zurich, Switz., Orell Fussli Publishing, 2000).

43. Thomas J. Crowley and Thomas S. Crowley, "How Warm Was the Medieval Warm Period?" *AMBIO: A Journal of the Human Environment* 29, no. 1 (2000): 51–54; Malcolm K. Hughes and Henry F. Diaz, "Was There a Medieval Warm Period, and If So, Where and When?" *Climate Change* 26, nos. 2–3 (1994): 109–42.

44. Ricardo Villalba, "Tree Ring and Glacial Evidence for the Medieval Warm Epoch and the Little Ice Age in Southern South America," *Climate Change* 26, nos. 2–3 (1994): 183–97; Perry and Hsu, "Geophysical, Archeological, and Historical Evidence Support a Solar-Output Model for Climate Change."

45. Barry A. Vann, *In Search of Ulster-Scots Land: The Birth and Geotheological Imagings of a Transatlantic People, 1603–1703* (Columbia: University of South Carolina Press, 2008).

46. The dates of occupation for Skara Brae and Mount Sandel fit into the warm periods identified by Perry and Hsu, "Geophysical, Archeological, and Historical Evidence Support a Solar-Output Model for Climate Change," pp. 12435–36.

47. Keigwin, "Little Ice Age."

48. Villalba, "Tree Ring and Glacial Evidence in the Medieval Warm Epoch and the Little Ice Age in Southern South America," p. 183.

49. Erica J. Hendy et al., "Abrupt Decrease in Tropical Pacific Sea Surface Salinity at End of Little Ice Age," *Science* 295, no. 5559 (2002): 1511–14.

50. Owen, *Natural Resource Conservation*, p. 579.

51. Ibid.

52. United States Geological Survey, "Cataclysmic 1991 Eruption of Mount Pinatubo, Philippines," http://pubs.usgs.gov/fs/1997/fs113-97/ (accessed January 12, 2011).

53. Ibid.

54. R. B. Symonds et al., in M. R. Carroll and J. R. Holloway, eds., "Volatiles in Magmas," *Reviews in Mineralogy* 30 (1994): 1–66.

55. Met Office, "Record Cold December 2010," http://www.metoffice .gov.uk/ (accessed January 13, 2011).

56. Ibid.

57. Hulme, "Conquering of Climate," p. 5.

CHAPTER 11. THE EAST'S HUMAN TSUNAMI

1. C. Alan Joyce et al., eds., *World Almanac and Book of Facts 2009* (New York: World Almanac Education Group, 2009), pp. 773–74.

2. Population data, including projections, for this section are taken from Jerome Fellman, Arthur Getis, and Judith Getis, *Human Geography: Landscapes of Human Activities* (Boston: McGraw Hill, 2007), pp. 503–508.

3. Bruce Crumley, "French Protests: The Cost of Sarkozy's Show of Force," *Time*, October 22, 2010, http://www.time.com/time/world/article/0,8599 ,2027077,00.html?xid=rss-mostpopular/ (accessed October 23, 2010).

4. William A. Schwab, *The Sociology of Cities* (Englewood Cliffs, NJ: Prentice Hall, 1992), p. 388.

5. Ibid.

6. Robert Famighetti, ed., *World Almanac and Book of Facts 1995* (Mahwah, NJ: Funk and Wagnalls, 1995), p. 767.

7. William A. McGeveran Jr., ed., *World Almanac and Book of Facts 2006* (New York: World Almanac Education Group, 2006), p. 779.

8. Office of National Statistics, "Census 2001: London," London, UK.

9. Simon Naylor and James R. Ryan, "The Mosque in the Suburbs: Negotiating Religion and Ethnicity in South London," *Social and Cultural Geography* 3, no. 1 (2002): 39–50.

10. The step-wise migration process is discussed at length by Dennis Conway, "Step-Wise Migration: Toward a Clarification of the Mechanism," *International Migration Review* 14, no. 1 (1980): 3–14.

11. Rich Morin, ed., "Muslim Americans: Middle Class and Mostly Mainstream," (Washington, DC: Pew Research Center, May 2007), http://pewresearch.org/pubs/483/muslim-americans/ (accessed July 5, 2011). For an estimate close to five million see McGeveran, *World Almanac 2006*, p. 714.

12. Barry A. Kosmin, Egon Mayer, and Ariela Keysar, *American Religious Identification Survey 2001* (New York: Graduate Center of the City University of New York, 2001), p. 13.

13. Surah 4:97.

14. Surah 4:100.

15. Surah 16:110.

16. As these differing viewpoints illustrate, there is a persisting divide between historians and geographers. James Rubenstein, *The Cultural Landscape: An Introduction to Human Geography*, 7th ed. (Upper Saddle River, NJ: Prentice Hall, 2002); Avihu Zakai, *Exile and Kingdom: History and Apocalypse in Puritan Migration to America* (Cambridge, UK: Cambridge University Press, 1992); Kerby Miller, *Irish Immigrants in the Land of Canaan: Letters and Memoirs from Colonial and Revolutionary America, 1675–1815* (Oxford: Oxford University Press, 2003).

17. Jacqueline Beaujeu-Garnier, "The Contribution of Geography" (published address, Ibadan University College, Ibadan, Nigeria, 1952) cited in "History, Geography, and the Still Greater Mystery of Historical Geography," ed. Chris Philo, in *Human Geography: Society, Space, and Social Science*, ed. Derek Gregory, Ron Martin, and Graham Smith (Minneapolis: University of Minnesota Press, 1994), p. 255.

18. For an excellent summary of his "laws," see Fellmann, Getis, and Getis, *Human Geography*, pp. 86–88.

19. Ibid., p. 87.

20. Ibid., p. 89.

21. For a discussion of the relationship between intended stay and return

migration, see Brigitte Waldorf, "Determinants of International Return Migration Intentions," *Professional Geographer* 47, no. 2 (1995): 125–36. To read about trans–Irish Sea migrations in the seventeenth century, see also my work titled "Presbyterian Social Ties and Mobility in the Irish Sea Culture Area," *Journal of Historical Sociology* 18, no. 3 (2005): 227–54.

22. Nil Demet Güngör and Aysit Tansel, *Determinants of Return Intentions of Turkish Students and Professionals Residing Abroad: An Empirical Investigation* (Bonn, Ger.: Institute for the Study of Labor, 2005).

23. Surah 4:100.

24. Surah 9:100.

25. The great Jewish Diaspora of the first century was not a voluntary migration into Europe. However, migrations in the 1800 and 1900s were certainly for political and economic reasons. The pogroms in czarist Russia and the genocidal assault inflicted on them by Nazi Germany were certainly reason enough to flee. Christians have historically found real and imagined political persecution handy reasons for escaping poor economic conditions. This is seen in the way they have sacralized "promised lands" while they de-sacralized their home lands. See my work titled *In Search of Ulster-Scots Land: The Birth and Geotheological Imagings of a Transatlantic People, 1603–1703* (Columbia: University of South Carolina Press, 2008).

26. Wilbur Zelinsky, "The Hypothesis of Mobility Transition," *Geographical Review* 61, no. 2 (1971): 219–49.

27. Huw Jones, "Evolution of Scottish Migration Patterns: A Social-Relations-of-Productions Approach," *Scottish Geographical Magazine* 102 (1986):151–64.

28. Barry A. Vann, *Puritan Islam: The Geoexpansion of the Muslim World* (Amherst, NY: Prometheus Books, 2011).

29. Max Weber, *Protestant Ethic and the Spirit of Capitalism* (New York: Scribner, 1958).

30. Population projections for this section are taken from Fellman, Getis, and Getis, *Human Geography*, pp. 503–508.

31. Vann, *Puritan Islam*.

32. Doubling time is calculated by using the Rule of 72 where 72 is the numerator and the Natural Increase Rate (NIR) is the denominator. The 2009 data are generated from C. Alan Joyce et al., eds., *World Almanac and Book of Facts 2009* (Pleasantville. NY: World Almanac Education Group, 2009), pp. 729–835.

33. Tourism in 2003 accounted for $1.016 billion income for Lebanon (5%

of GDP) and $402 million (30% of GDP) in Maldives during 2002. See McGeveran, *World Almanac 2006*, pp. 798, 802. In the case of Maldives, its economy depends heavily on immigrants from surrounding poor countries.

34. All of the data in this section, unless otherwise stated, are taken from Joyce et al., *World Almanac 2009*, pp. 729–835.

35. Ibid.

36. Douglas Cumming, Grant Fleming, and Armin Schwienbacher, "Legality and Venture Capital Exits," *Journal of Corporate Finance* 12, no. 2 (2006): 214–46.

37. Leslie S. Hiraoka, "Japan's Coordinated Technology Transfer and Direct Investments in Asia," *International Journal of Technology Management* 10, nos. 7–8 (2009): 714–31.

38. Steven Radalet and Jeffrey Sachs, "Asia's Reemergence: Capitalism Leaves in Western Enclave," *Foreign Affairs* 76, no. 6 (1997): 44–59.

39. Monty Guild and Tony Danaher, "Liquidity Flowing into Asia and Latin America," *Investment News: Leading News Source for Financial Advisors*, October 1, 2010, http://www.investmentnews.com/article/20101001/FREE/1010099 94/ (accessed February 18, 2011).

40. Economic and Social Commission for Asia and the Pacific, *Cities and Sustainable Development: Lessons and Experiences from Asia and the Pacific* (New York: United Nations, 2004), p. 77.

41. Ibid., p. 78.

42. Ibid.

CHAPTER 12. THE AMERICAS IN 2060

1. Alan Taylor, *American Colonies: The Settling of North America*, vol. 1 (New York: Penguin Books, 2002), p. 40.

2. C. Alan Joyce et al., eds., *World Almanac and Book of Facts 2009* (Pleasantville, NY: World Almanac Education Group, 2009), pp. 729–835.

3. These data were taken from United States Census Bureau, "Year of Entry of the Foreign-Born Population: 2009," (October 2010), http://www.census .gov/prod/2010pubs/acsbr09-17.pdf/ (accessed April 5, 2011); United States Census Bureau, "Place of Birth of Foreign-Born Population: 2009," http://usgovinfo .about.com/gi/o.htm?zi=1/XJ&zTi=1&sdn=usgovinfo&cdn=newsissues&tm

=23&gps=352_287_1090_450&f=00&su=p284.9.336.ip_&tt=2
&bt=0&bts=0&zu=http%3A//www.census.gov/population/www/socdemo/for
eign/index.html/ (accessed April 5, 2011).

4. Joyce et al., *World Almanac 2009*, pp. 550, 554.

5. Alan Strahler and Arthur Strahler, *Introducing Physical Geography*, 3rd
ed. (New York: John Wiley and Sons, 2003), p. 448.

6. Ibid.

BIBLIOGRAPHY

Akin, Wallace. *Forgotten Storm: The Great Tri-State Tornado of 1825* (Guilford, CT: Lyons Press, 2002).

"Annual Climatological Data." National Climate Data Center, US Department of Commerce. Cited in *World Almanac and Book of Facts 2009*, eds., Joyce et al.

A River Runs through It. Directed by Robert Redford. Burbank, CA: Tri-Star Motion Pictures, 1993. DVD.

Asian Development Bank. *Rise of Asia's Middle Class*. Manila, Phil.: Asian Development Bank, 2010.

Associated Press. "South Struggles to Rebound from Deadliest Tornado Outbreak since Great Depression." Fox News, April 29, 2011, http://www.foxnews .com/us/2011/04/29/survivors-picking-pieces-deadly-twisters-kill-27/.

Baird, Callicott, J., ed. *Companion to A Sand County Almanac: Interpretive and Critical Essays*. Madison: University of Wisconsin Press, 1987.

Baker, Alan R. H. *Geography and History: Bridging the Divide*. Cambridge, UK: Cambridge University Press, 2003.

Barrows, Harlan Read. "Geography as Human Ecology." *Annals of the Association of American Geographers* 13, no. 1 (1923): 1–14.

Beevy, E. S. "The Human Population." *Scientific American* 203, no. 3 (1960): 195–205.

Black, Henry Campbell. *Black's Law Dictionary*. 6th ed. Saint Paul, Minnesota: West Publishing, 1990.

Blaustein, Richard J. "Kudzu's Invasion into Southern United States Life and Culture." United States Department of Agriculture (2001). http://www.srs .fs.usda.gov/pubs/ja/ja_blaustein001.pdf/.

Bowen, E. G. *Britain and the Western Seaways*. London: Thames and Hudson, 1972.

———. "Le Pays de Galles," *Transactions and Papers of the Institute of British Geographers* 26 (1959): 1–23.

———. *Saints, Seaways, and Settlements in Celtic Lands*. Cardiff: University of Wales Press, 1969.

Boyer, Paul, et al. *The Enduring Vision: A History of the American People from 1865*. 5th ed. Boston: Houghton Mifflin, 2004.

Bradshaw, Michael, George W. White, Joseph P. Dymond, and Elizabeth Chacko. *Contemporary World Regional Geography: Global Connections, Local Voices*. 3rd ed. Boston: McGraw-Hill, 2009.

Braidwood, R. "From Cave to Village." In Lamberg-Karlovsky, ed., *Old World Archeology: Foundations of Civilizations*.

Brooks, Harold E., and Charles A. Doswell III. "Normalized Damage from Major Tornadoes in the United States: 1890–1999." *Weather and Forecasting* 16, no. 1 (2001): 168–76.

Burney, Charles. *Ancient Near East*. Ithaca, NY: Cornell University Press, 1977.

Buchsbaum, Ralph. *Animals without Backbones: An Introduction to the Invertebrates*. Chicago University of Chicago Press, 1948.

Carr, Donald. *Breath of Life*. New York: W. W. Norton, 1961.

Carson, Rachel. *Silent Spring*. Boston: Houghton Mifflin, 1962.

Caudill, Harry M. "Kentucky and Wales: Was Ellen Churchill Semple Wrong?" In McWhiney, *Cracker Culture: Celtic Ways in the Old South*.

———. *Night Comes to the Cumberlands: A Biography of a Depressed Area*. Boston: Little, Brown, 1963.

Chenoweth, Michael. "A Reassessment of Historical Atlantic Basin Tropical Cyclone Activity, 1700–1855." National Oceanic and Atmospheric Administration [NOAA] (2006). http://www.aoml.noaa.gov/ hrd/hurdat/Chenoweth/chenoweth06.pdf/.

Childe, Gordon. "The Urban Revolution." *Town Planning Review* 21, no. 1 (1950): 3–17.

Committee on Seismology. *Seismology: Responsibilities and Requirements of a Growing Science*. Washington, DC: Division of Earth Sciences National Academy of Sciences, 1969.

Conway, Dennis. "Step-Wise Migration: Toward a Clarification of the Mechanism." *International Migration Review* 14, no. 1 (1980): 3–14.

Crowley, Thomas J., and Thomas S. Crowley. "How Warm Was the Medieval Warm Period?" *AMBIO: A Journal of the Human Environment* 29, no. 1 (2000): 51–54.

Cumming, Douglas, Grant Fleming, and Armin Schwienbacher, "Legality and Venture Capital Exits." *Journal of Corporate Finance* 12, no. 2 (2006): 214–46.

Darby, H. C. *Relations of History and Geography: Studies in England, France, and the United States.* Exeter, UK: University of Exeter Press, 2002.

Davidson, Andrew Bruce. *The Book of Job.* Cambridge, UK: Cambridge University Press, 1889.

Davidson, Donald. *I'll Take My Stand: The South and the Agrarian Tradition, 75th Anniversary Edition.* Baton Rouge: Louisiana State University Press, 2006. First published 1930 by Harper.

DeCapua, Sarah. *The Shawnee.* Tarrytown, NY: Marshall Cavendish, 2006.

De Paor, Maire, and Liam De Paor. *Early Christian Ireland.* New York: Praeger, 1958.

Dickinson, Mary B. *Wonders of the Ancient World: National Geographic Atlas of Archeology.* Washington, DC: National Geographic Society, 1994.

Donovan, James. *A Terrible Glory: Custer and the Battle of Little Bighorn.* Boston: Little, Brown, 2008.

Du Bois, F. "The Cecchi Seismograph: A New Model Constructed for the Observatory of Manila." *Transactions of the Seismological Society of Japan* 5–8 (1885): 90–94.

Dubos, Rene. *Man Adapting.* New Haven: Yale University Press, 1969.

Durkheim, Émile, Carol Cosman, and Mark Sydney Cladis, *The Elementary Forms of Religious Life.* Oxford: Oxford University Press, 2002.

Economic and Social Commission for Asia and the Pacific. *Cities and Sustainable Development: Lessons and Experiences from Asia and the Pacific.* New York: United Nations, 2004.

Eddy, Don. "Up from the Dust." *Reader's Digest* 37 (1940): 20–22.

Ehrlich, Paul R. *Population Bomb.* New York: Ballantine Books, 1968.

Ehrlich, Paul R., and Ann Ehrlich. *Extinction: The Causes and Consequences of the Disappearance of Species.* New York: Random House, 1981.

———. *One With Nineveh: Politics, Consumption, and the Human Future.* Washington, DC: Island Press, 2004.

———. *Population Explosion.* New York: Simon and Schuster, 1990.

———. *The Dominant Animal: Human Evolution and the Environment.* Washington, DC: Island Press, 2008.

Eller, Ronald D. *Miners, Millhands, and Mountaineers: Industrialization of the Appalachian South, 1880–1930.* Knoxville: University of Tennessee Press, 1982.

Famighetti, Robert, ed. *World Almanac and Book of Facts 1995*. Mahwah, NJ: Funk and Wagnalls, 1995.

Feldman, Jay. *When the Mississippi Ran Backwards: Empire, Intrigue, Murder, and the New Madrid Earthquakes*. New York: Free Press, 2005.

Felknor, Peter. *Tri-State Tornado: The Story of America's Greatest Tornado Disaster*. New York: iUniverse, 2004.

Fellmann, Jerome D., Arthur Getis, and Judith Getis, *Human Geography: Landscapes of Human Activities*. 9th ed. Boston: McGraw Hill, 2007.

Firey, Walter. *Land Use in Central Boston*. Cambridge, MA: Harvard University Press, 1947.

Flynt, Wayne J. *Dixie's Forgotten People: The South's Poor Whites*. Bloomington: Indiana University Press, 1980.

Ford, Paul Leicester, ed. "Jefferson to Gallatin," September 8, 1816, *The Writings of Thomas Jefferson*. vol. 10. New York, 1892–1899, pp. 64–65. In C. Edward Skeen. *Journal of the Early Republic* 1, no. 1 (1981): 51–67.

Gans, Herbert J. *Urban Villagers: Group and Class in the Life of Italian Americans*. New York: Free Press, 1962.

Gaudette, Helen A., and G. A. Clark in Janssen, *World Almanac and Book of Facts 2008*.

Glantz, Michael H. *Societal Responses to Regional Climate Change: Forecasting by Analogy*. Boulder, CO: Westview Press, 1988.

Goldfield, David, et al. *The American Journey: A History of the United States*. Upper Saddle River, NJ: Pearson Prentice Hall, 2006.

Gonzales, Guillermo, and Jay Wesley Richards. *The Privileged Planet: How Our Place in the Cosmos Is Designed for Discovery*. Washington, DC: Regnery Press, 2004.

Grazulis, Thomas P. *Tornado: Nature's Ultimate Windstorm*. Norman: University of Oklahoma Press, 2001.

Gregory, Derek, Ron Martin, and Graham Smith. *Human Geography: Society, Space, and Social Science*. Minneapolis: University of Minnesota Press, 1994.

Gregg, Sara M. "Uncovering the Subsistence Economy in the Twentieth-Century South: Blue Ridge Mountain Farms." *Agricultural History Society* 78, no. 4 (2004): 417–37.

Grether, Jean-Marie, Nicole A. Mathys, and Jaime de Melo. "Global Manufacturing SO_2 Emissions: Does Trade Matter?" Kiel Institute, Hindenburgufer, Ger. http://www.wto.org/english/res_e/reser_e/gtdw_e/wkshop08_e/mathys _e.pdf/.

Güngör, Nil Demet, and Aysit Tansel. *Determinants of Return Intentions of Turkish Students and Professionals Residing Abroad: An Empirical Investigation*. Bonn, Ger.: Institute for the Study of Labor, 2005.

Hawley, Amos. *Human Ecology: A Theory of Community Structure*. New York: Ronald Press, 1950.

Hendy, Erica J., Michael K. Gagan, Chantal A. Alibert, Malcolm T. McCulloch, Janice M. Lough, and Peter J. Isdale. "Abrupt Decrease in Tropical Pacific Sea Surface Salinity at End of Little Ice Age." *Science* 295, no. 5559 (2002): 1511–14.

Hiraoka, Leslie S. "Japan's Coordinated Technology Transfer and Direct Investments in Asia," *International Journal of Technology Management* 10, nos. 7–8 (2009): 714–31.

Hirsch, Nathaniel. "An Experimental Study of the East Kentucky Mountaineers." *Genetic Psychology Monographs* 3 (March 1928): 183–244.

Hsu, Kenneth J. *Climate and Peoples: A Theory of History*. Zurich, Switz.: Orell Fussli Publishing, 2000.

Hubbard, Preston J. *Origins of the TVA: The Muscle Shoals Controversy, 1920–1932*. New York: W. W. Norton, 1961.

Hughes, Malcolm K., and Henry F. Diaz, "Was There a Medieval Warm Period, and If So, Where and When?" *Climate Change* 26, nos. 2–3 (1994): 109–42.

Hulme, Mike. "The Conquering of Climate: Discourses of Fear and Their Dissolution." *Geographical Journal* 174, no. 1 (March, 2002): 5–16.

Huntington, Ellsworth. "Geographer and History." *Geographical Journal* 43, no. 1 (1914): 19–32.

Huntington, Ellsworth, and Barry A. Vann, *Geography toward History: Studies in the Mediterranean Basin, Mesopotamia and Central Asia*. Piscataway, NJ: Gorgias Press, 2008.

Ingold, Tim. *The Perception of the Environment: Essays on Livelihood, Dwelling and Skill*. New York: Routledge, 2000.

Ivanko, John, and Lisa Kivirist. *Rural Renaissance: Renewing the Quest for the Good Life*. Gabriola Island, BC: New Society Publishers, 2009.

James, Preston E., and Geoffrey J. Martin. *All Possible Worlds: A History of Geographical Ideas*. 2nd ed. New York: John Wiley, 1981.

Jonassen, C. T. "Cultural Variables in the Ecology of an Ethnic Group." *American Sociological Review* 14, no. 5 (1949): 32–41.

Jones, Huw. "Evolution of Scottish Migration Patterns: A Social-Relations-of-Productions Approach." *Scottish Geographical Magazine* 102 (1986): 151–64.

Joyce, C. Alan, et al., eds., *World Almanac and Book of Facts 2009*. Pleasantville, NY: World Almanac Education Group, 2009.

Keigwin, Lloyd D. "The Little Ice Age and Medieval Warm Period in the Saragossa Sea." *Science* 274, no. 5292 (1996): 1503–1508.

Keun, Odette. *A Foreigner Looks at the TVA*. New York: Longmans, Green, 1937.

King, Duane H. *The Cherokee Indian Nation: A Troubled History*. Knoxville: University of Tennessee Press, 1979.

Kirshen, Paul, and Kelly Knee, and Mathias Ruth. "Climate Change and Coastal Flooding in Metro Boston: Impacts and Adaptation Strategies." *Climate Change* 90, no. 4 (2008): 453–73.

Knox, Paul L., and Sallie A. Marston, *Human Geography: Places and Regions in Global Context*. 4th ed. Upper Saddle River, NJ: Prentice Hall, 2007.

Kong, Lilly. "Geography and Religion: Trends and Prospects." *Progress in Human Geography* 14, no. 3 (1990): 355–71.

Kosmin, Barry A., Egon Mayer, and Ariela Keysar. *American Religious Identification Survey 2001*. New York: Graduate Center of the City University of New York, 2001.

Kuby, Michael, John Harner, and Patricia Gober. *Human Geography in Action*. Danvers, MA: John Wiley and Sons, 2004.

Lamberg-Karlovsky, Charles Clifford, ed. *Old World Archeology: Foundations of Civilizations*. San Francisco: W. H. Freeman, 1972.

Larson, Erik. *Isaac's Storm: A Man, a Time, and the Deadliest Hurricane in History* New York: Vintage Books, 1999.

Leighton, M. M. "Geology of Soil Drifting on the Great Plain." *Scientific Monthly* 47 (1938): 22–33.

Leopold, Aldo. *A Sand County Almanac: And Sketches Here and There*. Oxford: Oxford University Press, 1949.

Lewis, Howard R. *With Every Breath You Take*. New York: Crown, 1965.

Lichtenstein, Nelson, Susan Strasser, and Roy Rosenzweig. *Who Built America? Working People and the Nation's Economy, Politics, Culture, and Society since 1877*. New York: Worth, 2000.

Lion King, The. Directed by Roger Allers, and Rob Minkoff. Burbank, CA: Walt Disney Pictures, 1994.

Love, Paula McSpadden. "Clement Vann Rogers, 1839–1911." *Chronicles of Oklahoma* 48 (1970): 389–99.

Maher, Neil M. *Nature's New Deal: The Civilian Conservation Corps and the Roots of the American Environmental Movement*. Oxford: Oxford University Press, 2007.

Mallet, Robert. *Great Neapolitan Earthquake of 1857: The First Principles of Observational Seismology*. London: Chapman and Hall, 1862.

Malthus, Thomas. *An Essay on the Principle of Population*. Oxford: Oxford University Press, 1798.

Mather, Increase. *Remarkable Providences: An Essay for the Recording of Illustrious Providences*. Boston, 1674. Latest edition published in George Lincoln Burr, ed., *Narratives of the Witchcraft Cases 1648–1706*. New York: C. Scribner's Sons, 1914, pp. 3–38.

McArdle, Nancy. "Outward Bound: Decentralization of Population and Employment." Joint Center for Housing Studies, Harvard University, July 1999, http://www.jchs.harvard.edu/publications/communitydevelopment/mcardle_w99-5.pdf/.

McDonald, James R. "T. Theodore Fujita: His Contribution to Tornado Knowledge through Damage Documentation and the Fujita Scale." *Bulletin of the American Meteorological Society* 82, no. 1 (2001): 63–72.

McDonald, Michael J., and John Muldowny. *TVA and the Dispossessed*. Knoxville: University of Tennessee Press, 1982.

McGeveran, William A., Jr., ed. *World Almanac and Book of Facts 2003*. New York: World Almanac Education Group, 2003.

———, ed. *World Almanac and Book of Facts 2006*. New York: World Almanac Education Group, 2006.

McLean, Norman. *A River Runs through It, and Other Stories*. Chicago: University of Chicago Press, 1976.

McWhiney, Grady. *Cracker Culture: Celtic Ways in the Old South*. Tuscaloosa: University of Alabama Press, 1988.

Meyers, J. T. "The Origin of Agriculture: An Evaluation of Three Hypotheses." In Struever, *Prehistoric Agriculture*.

Miller, Kerby. *Irish Immigrants in the Land of Canaan: Letters and Memoirs from Colonial and Revolutionary America, 1675–1815*. Oxford: Oxford University Press, 2003.

Morin, Rich, ed. "Muslim Americans: Middle Class and Mostly Mainstream." Washington, DC: Pew Research Center, May 2007, http://pewresearch.org/pubs/483/muslim-americans/.

Morrison, Peter A., and Judith P. Wheeler, "Rural Renaissance in America? The

Revival of Population Growth in Remote Areas." Washington, DC: Population Reference Bureau, 1976.

National Geographic News, "The Deadliest Tsunami in History?" *National Geographic*, January 7, 2005, http://news.nationalgeographic.com/news/2004/12/1227_041226_tsunami.html/.

Naylor, Simon, and James R. Ryan. "The Mosque in the Suburbs: Negotiating Religion and Ethnicity in South London." *Social and Cultural Geography* 3, no. 1 (2002): 39–50.

Neuse, Steven M. *David Lilienthal: The Journey of an American Liberal.* Knoxville: University of Tennessee Press, 1996.

Newman, Marshall T. "The Blond Mandan: A Critical Review of an Old Problem." *Southwestern Journal of Anthropology* 6, no. 3 (1950): 255–72.

Nordhaus, William D., and Zili Yang. "A Regional Dynamic General-Equilibrium Model of Alternative Climate Change Strategies." *American Economic Review* 86, no. 4 (1996): 741–65.

Nuttli, Otto W. "The Mississippi Valley Earthquakes of 1811 and 1812: Intensities, Ground Motion and Magnitudes." *Bulletin of the Seismological Society of America* 63, no. 1 (1973): 227–48.

Office of National Statistics. "Census 2001: London." London, UK.

Oppenheimer, Clive. "Climatic, Environmental and Human Consequences of the Largest Known Historic Eruption: Tambora Volcano (Indonesia) 1815." *Progress in Physical Geography* 27, no. 2 (2003): 230–59.

Owen, Oliver S. *Natural Resource Conservation: An Ecological Approach.* 3rd ed. New York: Macmillan, 1980.

Owen, Oliver S., and Daniel D. Chiras. *Natural Resource Conservation: An Ecological Approach.* 5th ed. New York: Macmillan, 1990.

Pacione, M. "Relevance of Religion for a Relevant Human Geography." *Scottish Geographical Journal* 115 (1999): 117–31.

Park, Chris. *Sacred Worlds: An Introduction to Geography and Religion.* New York: Routledge, 1994.

Park, Robert E., Ernest W. Burgess, and R. D. McKenzie. *The City.* Chicago: University of Chicago Press, 1925.

Pattison, William D. "The Four Traditions of Geography." *Journal of Geography* 63, no. 4 (1964): 211–16.

Perry, Charles A., and Kenneth J. Hsu. "Geophysical, Archaeological, and Historical Evidence Support a Solar-Output Model for Climate Change." *Proceed-

ings of the National Academy of Sciences of the United States of America 97, no. 23 (2000): 12433–38.

Philo, Chris. "History, Geography, and the Still Greater Mystery of Historical Geography." In Gregory, Martin, and Smith, eds., *Human Geography: Society, Space, and Social Science.*

Poey, Andrés. "A Chronological Table, Comprising 400 Cyclonic Hurricanes Which Have Occurred in the West Indies in the North Atlantic within 362 Years, from 1493 to 1855." *Journal of the Royal Geographical Society* 25 (1855): 291–328.

Porter, Phillip W. "Geography as Human Ecology: A Decade of Progress in a Quarter Century." *American Behavioral Scientist* 22, no. 1 (1978): 15–39.

Prucha, Francis Paul. "Andrew Jackson's Indian Policy: A Reassessment." *Journal of American History* 56, no. 3 (1969): 527–39.

Radalet, Steven, and Jeffrey Sachs, "Asia's Reemergence: Capitalism Leaves in Western Enclave." *Foreign Affairs* 76, no. 6 (1997): 44–59.

Radmacher, Earl D., Ronald B. Allen, and H. Wayne House, eds. *New King James Study Bible.* 2nd ed. Nashville: Thomas Nelson, 2007.

Raines, John, ed. *Marx on Religion.* Philadelphia, PA: Temple University Press, 2002.

Ravenstein, E. G. "Birth Places of the People and the Laws of Migration." *Geographical Magazine* 3 (1876): 173–77, 201–206, 229–33.

———. "Laws of Migration." *Journal of the Statistical Society* 48, no. 2 (1885): 167–235.

Reardon, Thomas, C. Peter Timmer, Christopher B. Barrett, and Julio Berdequé. "The Rise of Supermarkets in Africa, Asia, and Latin America." *American Journal of Agricultural Economics* 85, no. 5 (2003): 1140–46.

Regier, H. A., and W. L. Hartman. "Lake Erie's Fish Community: 150 Years of Cultural Stress." *Science* 180 (1973): 1248–55.

Robinson, Jo. *Pasture Perfect: The Far-Reaching Benefits of Choosing Meat, Eggs, and Dairy Products from Grass-Fed Animals.* Vashon, WA: Vashon Island Press, 2004.

Roosevelt, Franklin D. *On Our Way.* New York: John Day, 1934.

Rubenstein, James. *The Cultural Landscape: An Introduction to Human Geography.* 7th ed. Upper Saddle River, NJ: Prentice Hall, 2002.

Sauer, Carl O. *Agricultural Origins and Dispersals.* New York: American Geographical Society, 1952.

———. "Foreword to Historical Geography." *Annals of the Association of American Geographers* 31, no. 1 (1941): 1–24.

Schama, Simon. *A History of Britain: At the Edge of the World? 3000 BC to AD 1603*. London: BBC Worldwide, 2002.

Scheihing, Mark H., and Hermann W. Pfefferkorn. "Taphonomy of Land Plants in the Orinoco Delta: A Model for the Incorporation of Plant Parts in Clastic Sediments of Late Carboniferous Age of Euramerica." *Review of Palaeobotany and Palynology* 41, nos. 3–4 (1984): 205–40.

Schnore, L. F. "Geography and Human Ecology." *Economic Geography* 37, no. 3 (1961): 207–17.

Schwab, William A. *The Sociology of Cities*. Englewood Cliffs, NJ: Prentice Hall, 1992.

Semple, Ellen Churchill. "The Anglo-Saxons of the Kentucky Mountains: A Study in Anthropogeography." *Geographical Journal*. (1901): 588–623.

Shapiro, Henry D. *Appalachia on Our Mind: The Southern Mountains and Mountaineers in the American Consciousness, 1870–1920*. Chapel Hill: University of North Carolina Press, 1978.

Shearer, Peter M. *Introduction to Seismology*. Cambridge, UK: Cambridge University Press, 1999.

Skeen, C. Edward. "The Year without a Summer: A Historical View." *Journal of the Early Republic* 1, no. 1 (1981): 51–67.

Slone, Verna Mae. *What My Heart Wants to Tell*. Boston: G. K. Hall, 1979.

Smith, Adam. *An Inquiry into the Nature and Causes of the Wealth of Nations*. London: Strahan and Cadell, London, 1776.

Smout, T. C. *A History of the Scottish People 1560–1830*. London: Collins, 1998.

Soon, Willie, and Steven H. Yaskell, "A Year Without a Summer: A Weak Solar Maximum, a Major Volcanic Eruption, and Possibly Even the Wobbling of the Sun Conspired to Make the Summer of 1816 One of the Most Miserable Ever Recorded." *Mercury* (May–June 2003): 13–22.

Steinbeck, John. *The Grapes of Wrath*. New York: Viking Press, 1939.

Stothers, Richard B. "The Great Tambora Eruption in 1815 and Its Aftermath." *Science* 224, no. 4654 (1984): 1191–98.

Strahler, Alan, and Arthur Strahler. *Introducing Physical Geography*, 3rd ed. Hoboken, NJ: John Wiley and Sons, 2003.

Struever, Stuart, ed. *Prehistoric Agriculture*. Garden City, NY: American Museum of Natural History, 1971.

Stump, Roger W. "The Geography of Religion: Introduction." *Journal of Cultural Geography* 7, no. 1 (1986): 1–3.

Sykes, Bryan. *Saxons, Vikings, and Celts: The Genetic Roots of Britain and Ireland*. New York: W. W. Norton, 2007.

———. *The Seven Daughters of Eve: The Science that Reveals our Genetic Ancestry*. London: Bantam Press, 2001.

Symonds, R. B., et al. In "Volatiles in Magmas." Ed. Carroll, M. R., and J. R. Holloway. *Reviews in Mineralogy* 30 (1994): 1–66.

Talbert, Roy. "Arthur E. Morgan's Ethical Code for the Tennessee Valley Authority." *Publications of the East Tennessee Historical Society* 40 (1968): 119–27.

———. *Human Engineer: A. E. Morgan and the Launching of the Tennessee Valley Authority*. Master's thesis., Vanderbilt University, 1967.

Taylor, Alan. *American Colonies: The Settling of North America*. vol. 1. New York: Penguin Books, 2002.

Thomas, Colin. "Landscape with Figures: In the Steps of E. G. Bowen." *Cambria* 12 (1985): 24.

Trenberth, Kenneth. "Climate: Uncertainty in Hurricanes and Global Warming." *Science* 308, no. 5729 (2005): 1753–54.

Tuan, Yi-Fu. *Morality and Imagination: Paradoxes of Progress*. Madison: University of Wisconsin Press, 1989.

Twister. Directed by Jan De Bont. Burbank, CA: Warner Brothers, 1996.

Udias, Agustin, and William Stauder. "The Jesuit Contribution to Seismology." *International Geophysics* 81, no. 1 (2002): 19–27.

United States Census Bureau. "Place of Birth of Foreign-Born Population: 2009." http://www.census.gov/prod/2010pubs/acsbr09-15.pdf/.

———. "Year of Entry of the Foreign-Born Population: 2009." (October 2010), http://www.census.gov/prod/2010pubs/acsbr09-17.pdf/.

Valins, Oliver. *Identity, Space, and Boundaries: Ultra-Orthodox Judaism in Contemporary Britain*. PhD diss., University of Glasgow, UK, 1999.

Vann, Barry A. *In Search of Ulster-Scots Land: The Birth and Geotheological Imagings of a Transatlantic People, 1603–1703*. Columbia: University of South Carolina Press, 2008.

———. "Presbyterian Social Ties and Mobility in the Irish Sea Culture Area." *Journal of Historical Sociology* 18, no. 3 (2005): 227–54.

———. *Puritan Islam: The Geoexpansion of the Muslim World*. Amherst, NY: Prometheus Books, 2011.

———. *Rediscovering the South's Celtic Heritage*. Johnson City, TN: n.p., 2004.

———. *"Space of Time or Distance of Place": Presbyterian Diffusion in South-Western Scotland and Ulster, 1603–1690*. Berlin: VDM Verlag Dr. Muller, 2008.

Villalba, Ricardo. "Tree Ring and Glacial Evidence for the Medieval Warm Epoch and the Little Ice Age in Southern South America." *Climate Change* 26, nos. 2–3 (1994): 183–97

Vosburgh, John. *Living with Your Land*. New York: Scribner, 1968.

Wakimoto, Roger M., and James W. Wilson, "Non-Super-Cell Tornadoes." *Journal of Atmospheric Sciences* 117 (1989): 1113–40.

Waldorf, Brigitte. "Determinants of International Return Migration Intentions." *Professional Geographer* 47, no. 2 (1995): 125–36.

Weber, Max. *Protestant Ethic and the Spirit of Capitalism*. New York: Scribner, 1958.

Weeks, John R. *Population: An Introduction to Issues and Concepts*, 8th ed. Belmont, CA: Wadsworth, 2002.

Wells, Spencer. *Deep Ancestry: Inside the Genographic Project*. Washington, DC: National Geographic Society, 2007.

———. *The Journey of Man: A Genetic Odyssey*. Princeton, NJ: Princeton University Press, 2002.

Wilhite, Donald. "The Ogallala Aquifer and Carbon Dioxide: Are Policy Responses Applicable?" In Glantz, *Societal Responses to Regional Climate Change*.

Williams, John Alexander. *Appalachia: A History*. Chapel Hill: University of North Carolina Press, 2002.

Wise, William. *Killer Smog*. Chicago: Rand McNally, 1969.

Wizard of Oz, The. Directed by Victor Fleming. Burbank, CA: Warner Brothers, 1939.

Worster, Donald. *Dust Bowl: The Southern Plains in the 1930s*. New York: Oxford University Press, 2004.

Wright, John K. "Notes on Early American Geopiety." *Human Nature in Geography: Fourteen Papers, 1925–1965*. Cambridge, MA: Harvard University Press, 1966.

———. "Terrae Incognitae: The Place of Imagination in Geography," *Annals of the Association of American Geographers* 37 (1947): 1–15.

Yunger, Sebastian. *Perfect Storm: A True Story of Men against the Sea*. New York: HarperCollins, 1997.

Zakai, Avihu. *Exile and Kingdom: History and Apocalypse in the Puritan Migration to America.* Cambridge, UK: Cambridge University Press, 2002.

Zelinsky, Wilbur. "The Hypothesis of Mobility Transition." *Geographical Review* 61, no. 2 (1971): 219–49.

Zimmerer, Karl S. "Human Geography and the New Ecology: The Prospect and Promise of Integration." *Annals of the Association of American Geographers* 84, no.1 (1994): 108–25.

INDEX